Beasts of Antiquity
STEM-BIRDS IN THE SOLNHOFEN LIMESTONE

Beasts of Antiquity
STEM BIRDS IN THE SOLNHOFEN LIMESTONE

Matthew P. Martyniuk

PAN AVES

Copyright © 2014 by Matthew P. Martyniuk

All rights reserved. This book or any portion thereof may not be reproduced or used in any manner whatsoever without the express written permission of the publisher except for the use of brief quotations in a book review or scholarly journal. Reuse of photographs with the Creative Commons Attribution Share-alike (CC-BY-SA) license must include attribution to the same author given in the caption and a reference to the same license number given in the caption.

First Printing: 2014

ISBN 978-0-9885965-5-9

Pan Aves
Vernon, NJ 07462

www.panaves.com

For Nova.

Frontispiece & cover: Rhamphorhynchus muensteri.

About the Author

Matthew P. Martyniuk is an illustrator and science educator specializing in Mesozoic birds and avian evolution. He has been drawing prehistoric flora and fauna since he first held a pencil, and became fascinated with the dinosaur/bird transition after discovering a copy of Gregory S. Paul's *Predatory Dinosaurs of the World* at his local library.

He is a founding member of "Wikiproject Dinosaurs", an initiative to generate and curate scientifically precise content for the online encyclopedia Wikipedia. His illustrations and diagrams have appeared in a variety of scientific journals, books, news articles, and television programs, including the *Journal of Zoology*, *Wired*, and the BBC. He publishes the paleontology blog *DinoGoss*, at dinogoss.blogspot.com, and his online portfolio and publications list can be viewed at henteeth.com.

More from Matthew P. Martyniuk:
A Field Guide to Mesozoic Birds and other Winged Dinosaurs
Pan Aves, 2012

Contents

Introduction: Crowns & Stems 1
Limestone & Lithography 4
The Solnhofen Ecosystem 13
The Solnhofen Stem-Birds 18
    -Pterosaurs 27
        "The Very Singular Skeleton" 32
        Interpreting the Pterodactyl 37
        Pterodactylus antiquus 44
        Gnathosaurus subulatus 48
        Ctenochasma elegans 52
        Rhamphorhynchus muensteri 56
        Scaphognathus crassirostris 62
        Germanodactylus rhamphastinus 66
        Ardeadactylus longicollum 70
        Cycnorhamphus suevicus 72
        Germanodactylus cristatus 76
        Anurognathus ammoni 80
    -Theropods 84
        Compsognathus longipes 86
        Archaeopteryx lithographica 90
        Archaeopteryx siemensii 94
Future Finds 100
Glossary ci
Taxonomic Index cii
Bibliography civ
Index cviii
Image Credits cxii

Introduction: Crowns & Stems

In 1735, the Swedish biologist Carl Linnæus, began the modern system of classification by publishing his book *Systema Naturae*. Linnæus attempted to place all life into the familiar, ranked groups of Kingdom, Phylum, Order, Genus, and Species, based on shared characteristics or "types". This kind of type-based (typological) system was fine for facilitating communication between biologists, but problems with the Linnaean system arose when 18th and 19th century scientists began to recognize that species were not static. The discovery that life evolves, and that there were species in the past which had since become extinct, created new challenges for scientists used to placing all species into separate boxes. Linnaean classification worked well for categorizing living species, but fossil forms showed that the dividing lines between these groups were blurry.

In the 1950s, a German biologist named Willi Hennig helped to start a new revolution in scientific classification. Hennig's system, called phylogenetics, was designed to group organisms by evolutionary relationships into nested groups called clades. Rather than having to draw arbitrary dividing lines between groups, scientists could now simply nest a group inside the group it had evolved from to illustrate their closer relationship relative to other groups.

Besides better integrating evolutionary biology into classification, this system also eventually helped to eliminate many unjustified assumptions about fossil species that

Previous page: Pages from Linnaeus' 1729 thesis, Praeludia Sponsaliorum Plantarum.

had been made based on typological thinking. Previously, many researchers would assume physical features or even behaviors about extinct animals based on their Linnaean category. *Archaeopteryx*, for example, were classified as birds

Painting of Carl Linnaeus by Hendrik Hollander, 1853.

(Class *Aves*), and so therefore must have had the essential traits we associate with all birds alive today, traits like flapping flight, fast growth, parental care, and "warm blood". Largely because of these typological assumptions, it has taken scientists over a century to begin truly understanding the real-life biology of ancient bird relatives like *Archaeopteryx*. When the evidence is examined critically, it turns out that *Archaeopteryx* and many other prehistoric "birds" may not have been able to attain true flight, and probably had a slower, more reptile-like growth pattern. In cases like this, simply classifying these animals as "birds" has contributed to a misunderstanding of the biology of ancient animals.

Using phylogentic classification can help challenge these kinds of unjustified assumptions. Phylogenetics classifies *Archaeopteryx* as a non-bird, because it is not a member of the avian crown group. In phylogenetics, a crown group is the natural group formed by all members of a certain lineage still alive in the present day. Crown group birds include all modern birds groups, from songbirds to ostriches and kiwis. It does not include any animals that are part of the overall bird lineage that branched off earlier than the common ancestor of songbirds and ostriches, because as far as we are able to tell, all such animals are now extinct. Any animals that are more closely related to modern birds than to the closest living relatives of birds, the crocodilians, are classified as part of the avian stem group. With this new classification system, it is possible to avoid assumptions about the biology of prehistoric species by splitting them into "crown-birds" and "stem-birds", rather than lumping them together as just "birds".

Exploring the biology and diversity of stem-birds, the long and fascinating lineage bridging ancient archosaurian ancestors with true modern birds, will be the focus of this book.

Simplified archosaur cladogram illustrating the relationship between panavians (blue), stem-birds (green), and birds (crown Aves, orange).

3

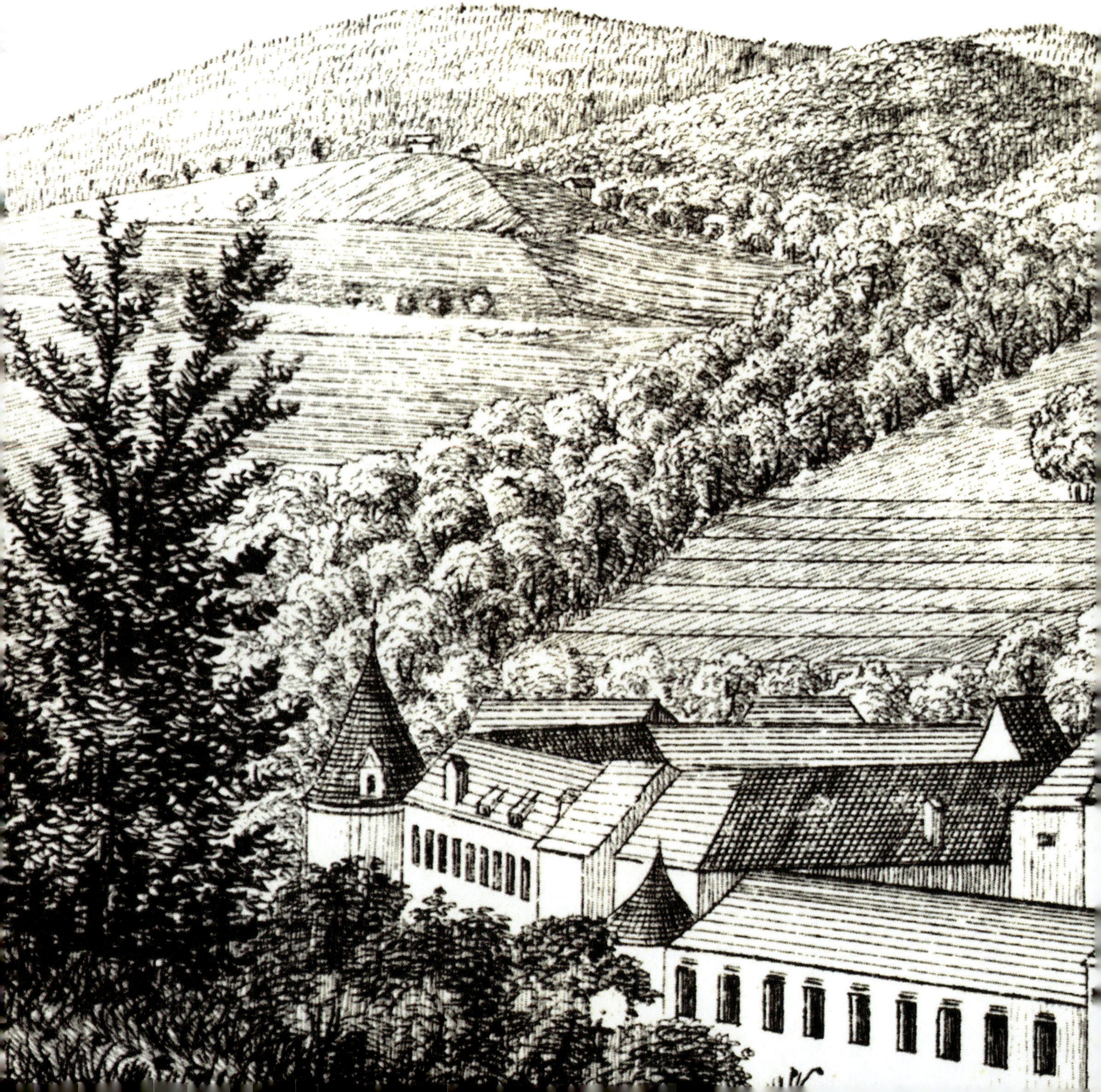

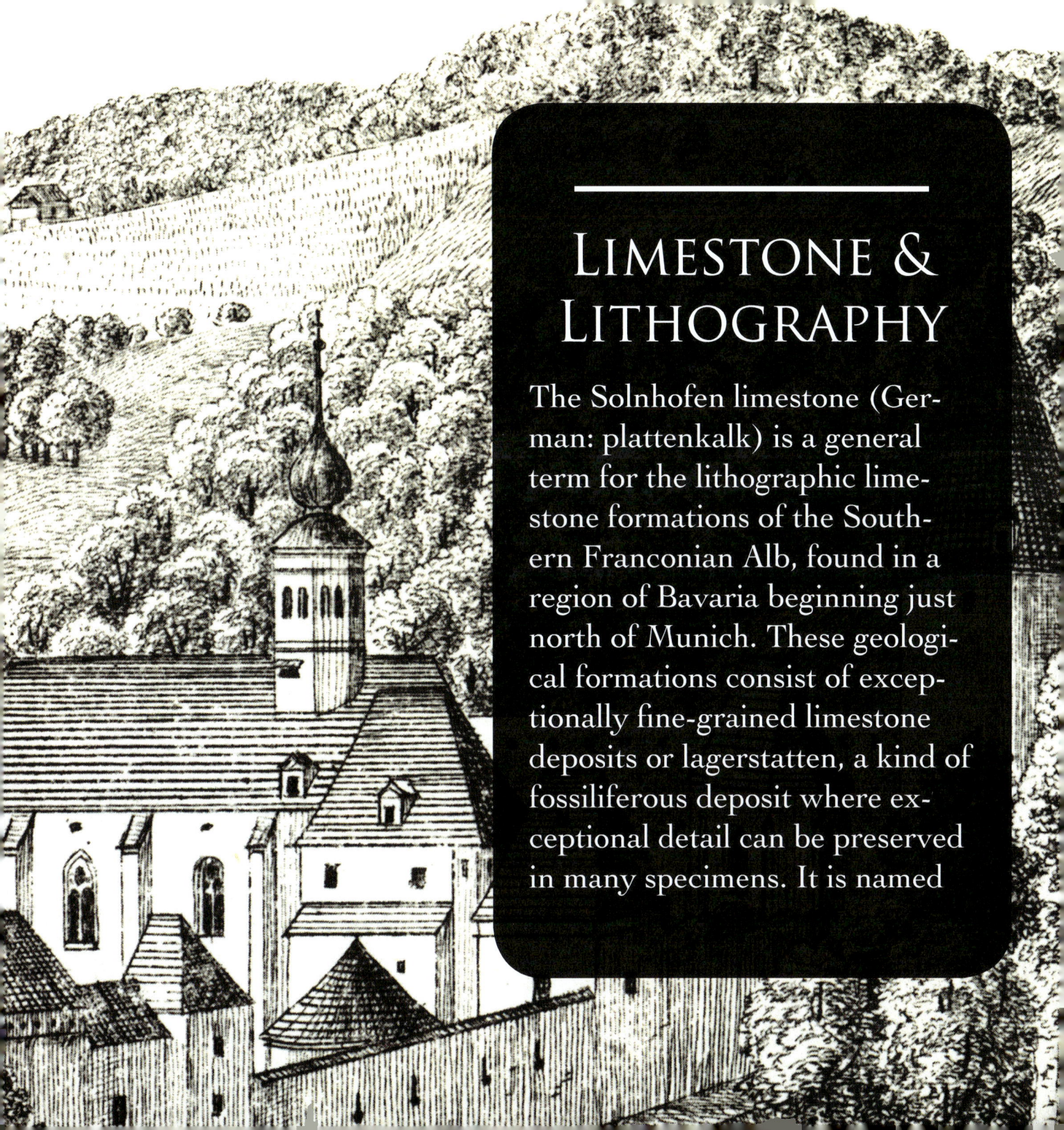

Limestone & Lithography

The Solnhofen limestone (German: plattenkalk) is a general term for the lithographic limestone formations of the Southern Franconian Alb, found in a region of Bavaria beginning just north of Munich. These geological formations consist of exceptionally fine-grained limestone deposits or lagerstatten, a kind of fossiliferous deposit where exceptional detail can be preserved in many specimens. It is named for the town of Solnhofen where much of the lithographic stone was quarried beginning in the late 18th century. Solnhofen itself is named for St. Sola, a monk sent to convert the local population to Christianity around 750 C.E.

The terms "Solnhofen limestones" and "Solnhofen archipelago" are sometimes used interchangeably to describe the fossil beds of the Solnhofen Limestone Formation themselves, as well as related formations both geologically higher and lower, in and around southern Germany and northern France (Wellnhofer, 2009). The Solnhofen Formation itself is best known from limestone quarries in Solnhofen, Langenaltheim, Kelheim (incl. Goldberg), Workerszell, Eichstätt, Zandt, Ried, Pfalzpaint, and Hienheim (Schweigert, 2007; Rauhut et al., 2012). Additional fossiliferous quarries contain both pre- and post-Solnhofen faunas, such as the underlying Painten Formation (sometimes considered part of a larger Rögling Formation) which preserves a similarly high diversity of stem-birds to that of the Solnhofen proper. Younger nearby strata preserve the fossils of the Moernsheim (Mörnsheim) Formation, which includes the Daiting quarry where a fragmentary stem-bird attributed to *Archaeopteryx* has been found. The top of the Moernsheim Formation has been dated at 150 million years old (Rauhut et al., 2012), but no direct dating of the lower portion of the Solnhofen limestone has been possible so far. The Painten Formation appears to span the boundary of the Kimmeridgian and Titho-

Steel engraving depicting Eichstätt, Johann Poppel / Heinrich Adam, 1880

A lithographic limestone quarry.

Portrait of Alois Senefelder, by Franz Hanfstaengl, 1854

nian stages of the Upper Jurassic, a boundary which has been dated elsewhere to 152 million years old. This means that the Solnhofen Formation itself is early Tithonian in age, and probably formed between about 151.5 and 150.8 million years ago.

These limestones of the Southern Franconian Alb were first used for carving and drawing during the Stone Age, and have been used as building materials for centuries, with evidence dating back to the forts of ancient Rome and medieval houses (Barthel, 1991). It was with the invention and widespread use of lithography that the exploitation of limestone quarries in the Alb, and the examination of the fossils found embedded in the stones, began to increase dramatically during the 19th century (Bottjer, 2002). Lithography is the art of printing using acid to etch relief images into flat, fine-grained limestone plates, to which ink can then be applied and used as a stamp to make prints. This process was invented by the German author Alois Senefelder in the 1790s. The limestone around Solnhofen had an exceptionally fine grain and porous surface compared even to other regions in the Alb, and became famous for its lithographic limestone. Stone in other areas had been, and continues to be, a popular local source of building materials, such as roof and floor tiles taken from the quarries at Eichstätt, and of crushed lime from the quarries at Painten (Barthel, 1991).

Fossils are relatively rare in the Solnhofen limestone, but quarrying has been going on for so long that their presence there has been well known since the limestone was first used (Bot-

A worker cuts limestone slabs in a Mörnsheim Formation quarry outside the town of Solnhofen.

7

Left: Fossil specimen of a shrimp (Aeger elegans) *from the Solnhofen limestone, in the collections of the Museum of Natural History, Berlin.*

tjer, 2002). Fossils from the limestone have been considered objects of great value or even as magical talismans since the Stone Age, and have been collected and sold by locals ever since (Barthel, 1991).

The rare quality of the Solnhofen limestone sediments is thought to be due to the manner in which these sediments were laid down on the ancient seafloor. During the late Jurassic period, this region of Bavaria was part of a shallow sea dotted with islands and peninsulas, many of which formed tranquil lagoons sheltered by sponge reefs. These sponge reefs towered over 50 meters above the seafloor in some areas, creating fully enclosed basins almost completely closed off to water flow.

Flow was closed off both between basins, from the waters to the north, and from the Tethys Sea to the south (Bottjer, 2002). During the time the limestone formations were deposited in the Tithonian age of the Late Jurassic, conditions were beginning to change, with the tops of the sponge reefs dying off. However, these were gradually replaced with coral, maintaining the walls between sheltered basins. This extremely long period of isolation caused the water in the basins to become stratified, with highly saline conditions at the bottom that prevented any animal life from surviving there. This is strikingly attested to by the large number of marine arthropod fossils found in the limestones, combined with the near absence of any tracks that would be expected in an active seafloor ecosystem (Barthel, 1970; Bottjer, 2002). In the rare instance that arthropod trackways are found, the track-maker is often found dead and fossilized at the end, as in the case of some fossil horseshoe crabs (a discovery that falsified the earlier hypothesis that these were pterodactyl tracks!). It is possible that the water at the basin floor was

A limestone quarry at Langenaltheim.

also oxygen deprived to some degree, though not completely anoxic, and bottom oxygen levels probably varied throughout the region and between different basins. There is no evidence that the bottom was completely anoxic, though it may have been nearly so in some areas (Bottjer, 2002). This near total absence of any water flow or animal activity on the basin floors allowed dissolved carbonate particles to very slowly settle and form an intensely fine-grained, undisturbed limestone. The sources of this carbonate are not completely understood, and there are several hypotheses for the exact method of sedimentation, though they are not necessarily mutually exclusive.

One source may have been the coral reefs which were present to the south of the archipelago. An "ooze" of carbonate-saturated water from the coral would have occasionally been stirred up by rough weather and periodic storms originating in the Tethys Sea, and washed towards the Solnhofen basins. Only the most miniscule suspended particles of carbonate would float high enough to be washed over the top of the sponge reef mounds that surrounded the basins, and would gradually have settled to the basin bottom. In this way, the sponge reefs may have acted as a filter, allowing only the finest limestone grains to sift

Fossil specimen of a Solnhofen dragonfly (Cymatophlebia longialata), in the collections of the Bürgermeister Müller Museum, Solnhofen.

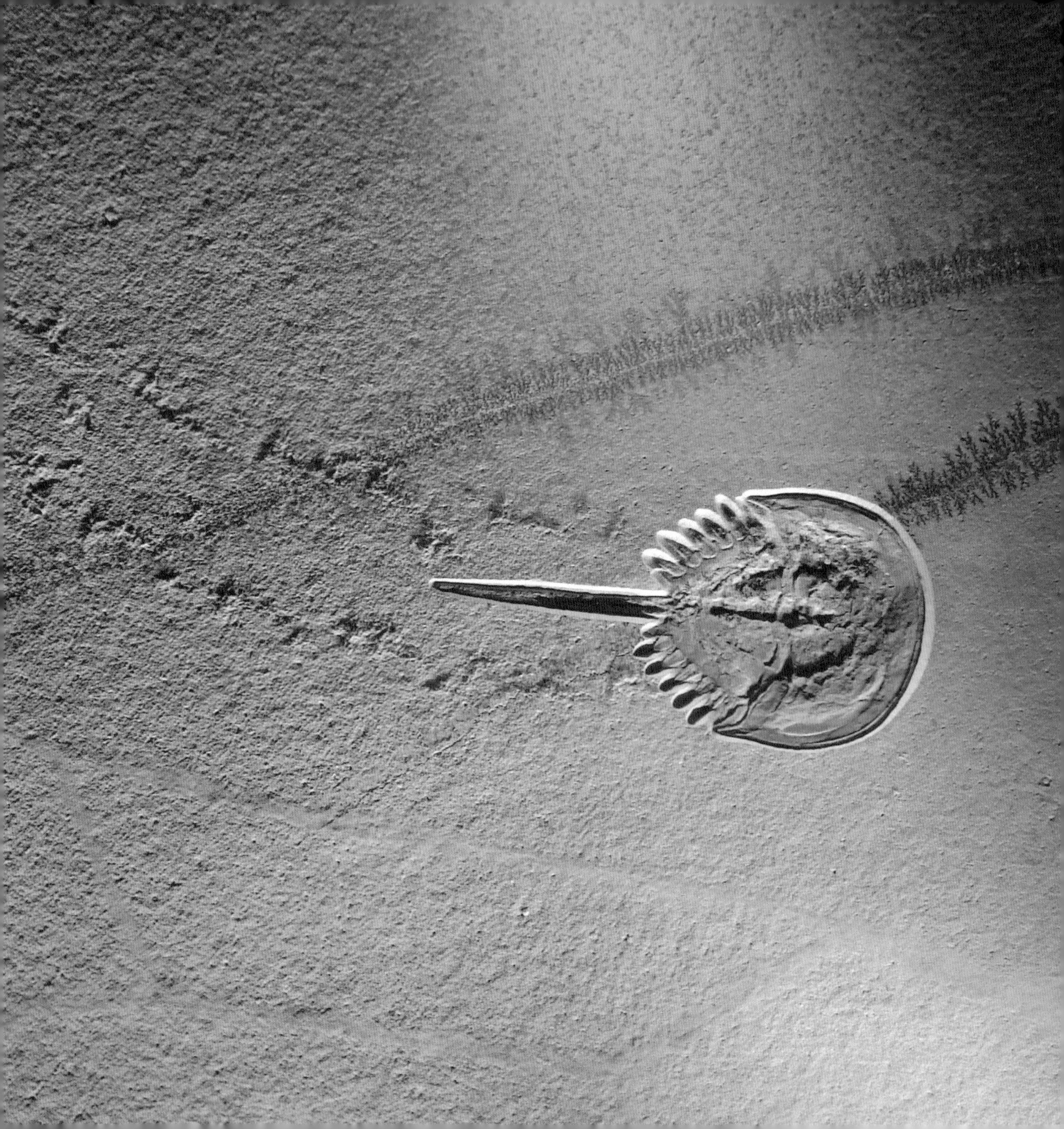

into their sheltered basins. It is only in the eastern part of the Alb near Painten, close to where coral was actively replacing the dying sponge reefs, that larger carbonate debris is found mixed in with the fine-grained limestone (Bottjer, 2002). Larger tropical storms and occasional typhoons would have thoroughly mixed open-ocean water into the basins, interrupting the limestone deposition process. This would explain the regular interspersion of marly, clay-filled layers (probably washed in from land by intense rain) found in between layers of fine-grained limestone. Mixing of cold, open-ocean water with warm waters of the basins would have also re-enriched them with carbonate, and begun the slow limestone precipitation and deposition process over again (Wellnhofer, 2009).

Another, possibly major, contributor to the limestone formation in the Solnhofen lagoon may have been the growth of cyanobacteria. This blue-green algae consists of tiny microorganisms surrounded by cal-

*Left: A fossil specimen of a horseshoe crab (*Mesolimulus walchi*) which died and was preserved in the process of leaving a trackway, probably as the result of the harsh conditions at the bottom of a Solnhofen basin it wandered into accidentally. Part of the collections of the Museum of Man and Nature, Munich.*

cium shells. These tiny spheres are present in large numbers throughout the limestone slabs. As blue-green algae colonies grow, they secrete/precipitate calcium, which accumulates in the seafloor sediment until disturbed. Periodic mixing of water from Tethys storms would have occasionally disturbed these algae colonies, which then re-grew, forming a finely layered pattern in the sedimentary rock. When dead animals and plants settled on the seafloor, the algae colonies would have covered them, quickly encasing them in a layer of carbonate, and preventing further decomposition. In effect, the entirety of the fine-grained Solnhofen limestone deposits may have actually been stromatolites, biochemical accretions formed by microbial mats (Wellnhofer, 2009).

Any animals that wound up on the basin floor would have quickly died from the hostile, highly saline and nearly anoxic conditions, and the lack of decomposers allowed time for the carbonate sediments to accumulate and bury their remains. The carcasses of animals that died near the surface would also have eventually found their way to the bottom of the basins and been similarly buried, with the fine grain of the sedimentary rock preserving any soft tissues that may have survived the journey.

The Solnhofen Ecosystem

During the late Jurassic, most of what is now central Europe consisted of an archipelago of large islands situated in a shallow extension of the southern Tethys Sea and the western Atlantic Ocean. At that time, this portion of Europe was much farther south than it is today, situated between about 60 degrees and 30 degrees north. Today, this region is occupied by the Mediterranean Sea. The limestone quarries of

Solnhofen and related sites would have been at the same latitude as the southern Greek islands and Cyprus are today. The Solnhofen lagoon, a region of the sea sheltered on three sides by large islands, contained a number of underwater basins isolated by mounds of built-up sponge reefs. These basins were interspersed with high reef mounds possibly forming small islets. The basins were sheltered by a large island to the north and west, called the "Mitteldeutsche" (central German) Landmass, and a smaller island, the Bohemian Landmass, to the east. To the south, a series of coral reefs sheltered the Solnhofen basins from the deeper waters of the Tethys Sea. Many local sponge reefs were dead or dying at the time the Solnhofen fauna were preserved in limestone, and were in the process of being replaced by coral at the tops of the mounds they had built up during the Middle Jurassic. Most marine life preserved in the fossil beds associated with the Southern Franconian Alb are shallow-water or reef-associated species, including cephalopods like nautilus and ammonites, rays, and bony fishes similar in form to modern reef fish. Shallow-water tetrapods including many species of turtle, a stem-tuatara (*Pleurosaurus goldfussi*), and several

Early illustrations of a Rhamphorhynchus *and an* Archaeopteryx *(above) and* Pterodactylus *(left) in their habitat, from The World Before the Deluge by L. Figuier, 1863. Note the inaccurate depiction of numerous tall trees near the shore.*

stem-crocodilians are also known (Bottjer, 2002), as are rare occurrences of open-water species like sharks, plesiosaurs (known only from vertebrae; Barthel, 1990), and at least one species of ichthyosaur (*Aegisaurus leptospondylus*). These non-reef related species probably washed into the basins during storms, possibly after already having died in many cases. Though evidence suggests the presence of extensive reefs, actual remains of reef-producing organisms are rare in these formations (Bottjer, 2002).

Remnants of land-dwelling life are, understandably, much rarer among the Solnhofen fossils. We don't know if the edges of the Solnhofen islands were mainly cliffs

NORTH SEA
Berlin
Szczecin
Mittleand Passage
London-Brabant Pen.
Mitteldeutsche Is.
Bohemian Is.
Amsterdam
Elbe Passage
40° N
London
N
Prague
Brussels
Luxembourg
Solnhofen Arch.
Eichstätt
Kelheim
Solnhofen
Paris
Linz
Coral Reef
Paris Basin
Vaduz
Berne
35° N
Centrale Is.
Ljubljana
Milan
TETHYS SEA
Canjuers
Monaco
300 km
300 mi

or sandy beaches, or some combination of the two. Whatever the topography, plant remains suggest a hot and arid climate, with a long dry season and very short rainy season. The coastal areas of the low-lying islands probably saw very little rainfall even during the "rainy" season, and would have resembled modern coastal deserts (Paul, 2010). Fossil leaves and pollen show that a desert-adapted flora lived near the coast, including shrub-like conifers, seed ferns, and cycadeoids. No fossil trees are known from this region, and no fossil tree pollen has ever been found. A single leaf similar to that of a *Ginkgo* was described as an algae frond in 1854, and later interpreted as a possible tree leaf, though it lacks veins characteristic of *Ginkgo* and the ginkgo-like shape is probably a coincidence (Wellnhofer, 2009). This lack of tree fossils suggests that plants taller than about three meters, or any plants with trunks, were either totally absent or very rare in the coastal areas of the Solnhofen archipelago. Even when present, trees were probably restricted to the interiors of the large islands to the north and east (Wellnhofer 2009; Paul, 2010). Insects were abundant and diverse on land, but known vertebrates were limited to several species of stem-lizards, stem-tuatara, stem-birds, and stem-crocodilians. Taken together, the evidence suggests a dry, subtropical climate, with high evaporation creating very salty waters. However, the presence of insects that require freshwater to reproduce, like mayflies, suggests the presence of at least seasonal freshwater streams near the coast (Bottjer, 2002), probably flowing to the sea from inland water sources.

Left: Map showing a reconstruction of north-central Europe during the late Jurassic period. Select modern cities are marked in white. The location of major Solnhofen Limestone quarries are marked in black.

Geography of island interiors mainly speculative. Small islands formed from the top of reefs in the shallowest portions of the sea would have come and gone relatively quickly over geological time, due to changes in sea level and erosion. The Mitteldeutsche and Bohemian land masses may have been joined together, at least occasionally.

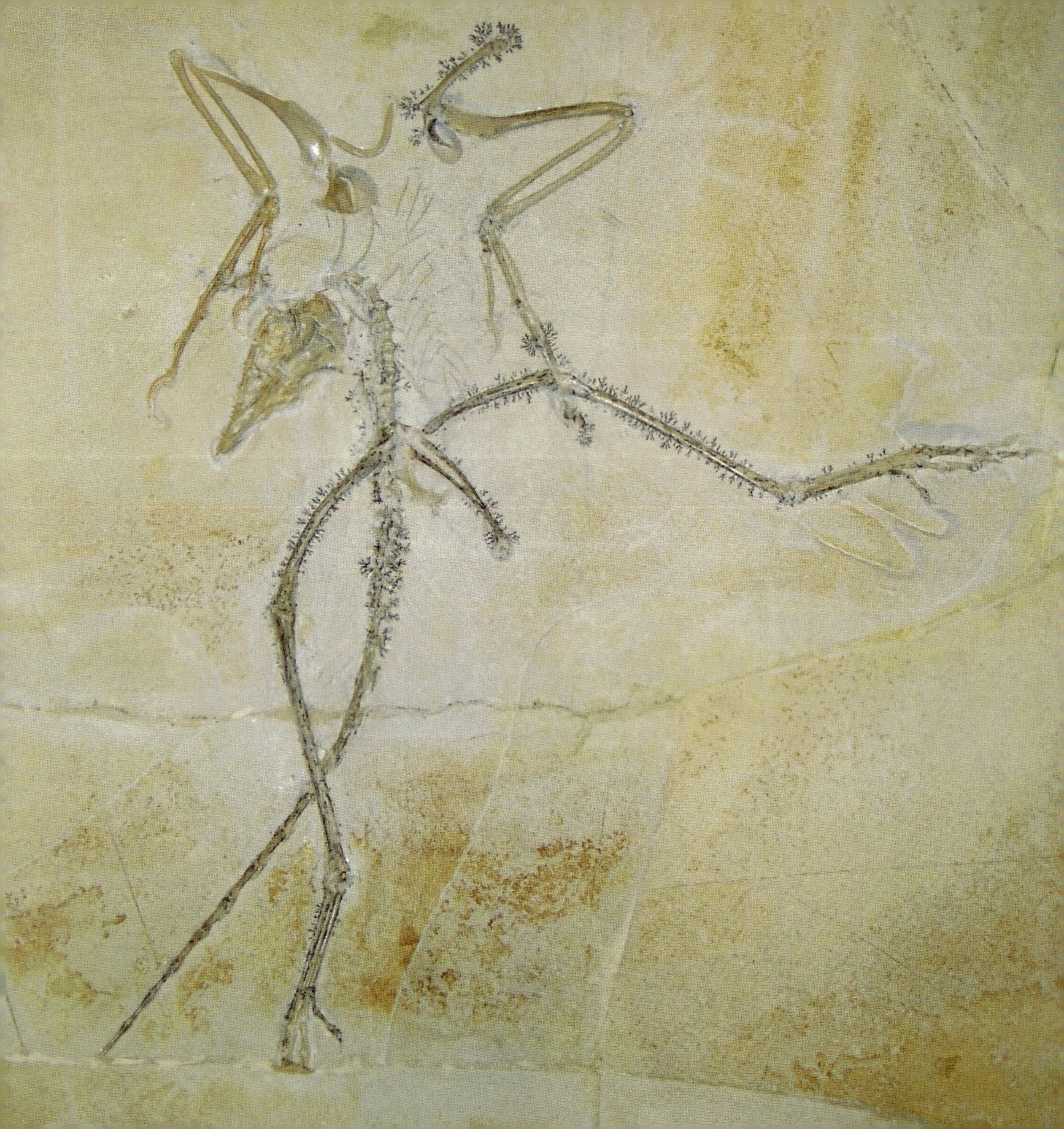

The Solnhofen Stem-Birds

The earliest stem-bird discoveries in the Solnhofen lithographic limestones were also coincidentally among the first fossils that helped scientists realize that extinct animals existed at all. Fossils had been known throughout human history, and were often interpreted as giants or dragons that had been driven away from civilization by mythic heroes, or even as organic patterns in rock formed by the life-generating force of the earth. It was

Life restoration of an adult Gnathosaurus subulatus *in profile.*

not until the birth of the science of comparative anatomy, in the late 18th century, that fossils were re-interpreted and re-evaluated as the remains of known animal species and their ancient relatives.

Among the first extinct species named by science were ice age mammals like the mastodon (as *Elephas americanus* in 1792), cave bear (*Ursus spelaeus*, 1796), and ground sloth (*Megatherium americanum*, 1796). These discoveries helped prove the hypothesis that the extinction of entire species was not just possible, but had actually occurred

Previous page: The Thermopolis specimen of Archaeopteryx siemensii.

in Earth's past. However, many of these species appeared to simply be strange or giant variations on modern species, rather than completely new lineages.

The first scientifically described stem-bird fossil from Solnhofen was, understandably, misinterpreted when first discovered.* A jumble of long, spindly bones lacking a skull, this tiny specimen made its way into the mineral collection of Archduchess Maria Anna of Austria some time after 1757, when she began collecting rocks and minerals as a hobby while recuperating from tuberculosis. The fossil, later known as the "Pester Exemplar", was first studied in 1779 by Romanian minerologist Ignaz von Born, who decided that it represented a fossil crustacean (Osi et al., 2010). After more complete specimens were recognized,

Portrait of Archduchess Maria Anna, owner of one of the first scientifically described pterosaur specimens. By Martin van Meytens, 18th c.

20

Life restoration of an adult Pterodactylus antiquus *in profile.*

scientists would later recognize this as a specimen of a juvenile pterosaur named *Pterodactylus micronyx*, a likely synonym of the species now known as *Gnathosaurus subulatus* (Bennett, 2013).

*Though it was the first Solnhofen stem-bird, this was not the first stem-bird to ever be scientifically described. That honor goes to the broken end of a megalosaurian thigh bone discovered in England and studied by Robert Plot in 1676, and which was later labelled *Scrotum humanum*. Plot thought this bone possibly belonged to a Roman war elephant. Additional stem-bird specimens, a sauropod tooth and a megalosaur tooth, were also found in England and described in 1699 by Edward Lhwyd, who gave them the names *Rutellum impicatum* and *Plectronites belemnitum*, respectively (Delair & Sarjeant, 2002; Evans, 2010).

It was a second specimen found at about the same time as the Pester Exemplar that kick-started the study of stem-birds. The first complete specimen of a stem-bird, the first given an official name, and in fact the first fossil specimen ever recognized as a member of a wholly extinct lineage, is known as the Mannheim Specimen. The early French anatomist Georges Cuvier named it the *"Ptero-Dactyle"*, later revised to the formal scientific name *Pterodactylus*. Soon after the Mannheim Specimen was described, a relationship with birds was suggested by some scientists. This remained controversial, however, and eventually the idea that these "pterodactyls" were reptiles gained consensus. Still, they were clearly a type of reptile not alive in the modern world, and this helped demonstrate that concepts of extinction and deep time were real. More pterodactyls, later called pterosaurs, were soon found, adding to their known diversity.

Several more major discoveries in stem-bird evolution were made in the Solnhofen and associated limestones in the mid-19th century. The first, unearthed in 1859, was described and given the name *Compsognathus longipes* two years later. While clearly sharing characteristics with reptiles, this was a small, bipedal creature which also had very bird-like features in its skeleton. In 1861, the same year *Compsognathus* was named, the first specimens of an animal clearly related to modern birds was found in the Solnhofen beds. Classified in the new fossil species *Archaeopteryx lithographica*, these were immediately classified as "birds" on the basis of feather impressions associated with their skeletons. In 1863, Karl Gegenbauer became among the first anatomists to note a similarity between the skeletons of *Archaeopteryx* and *Compsognathus*. Edward Drinker Cope and Thomas Henry Huxley later supported this view, with Huxley using the many similarities between the two as evidence in support of Darwin's new theory of evolution (Switek, 2010). In essence, Gegenbauer was the first to recognize both as members of the bird lineage. *Compsognathus longipes* could be viewed as the first widely accepted stem-bird species: a fossil reptile that shared more features with birds than with any other reptile species, living or extinct, known at the time.

To date, between about 13 and 17 species of stem-birds have been found in the Solnhofen Limestone Formation, depending on how species based on juvenile specimens are classified. In the early days of paleontology, it was assumed that stem-birds retained the same general proportions and features, like number of teeth, throughout their lives. This, combined with the fact that most Solnhofen stem-bird fossils are the remains of juveniles, led to numerous species being named for what are now understood to be growth stages of relatively fewer species. The genus *Pterodactylus*, for example, at one time contained over ten species from the Solnhofen limestones alone, not even including those based on specimens later reclassified in other genera. Better understanding of the life history of stem-birds, especially within

Next page: Restoration of a Solnhofen beach ecosystem. A: Archaeopteryx lithographica.
B: Pterodactylus antiquus. *C.* Rhamphorhynchus muensteri.
D. Compsognathus longipes. *E.* Ctenochasma elegans.

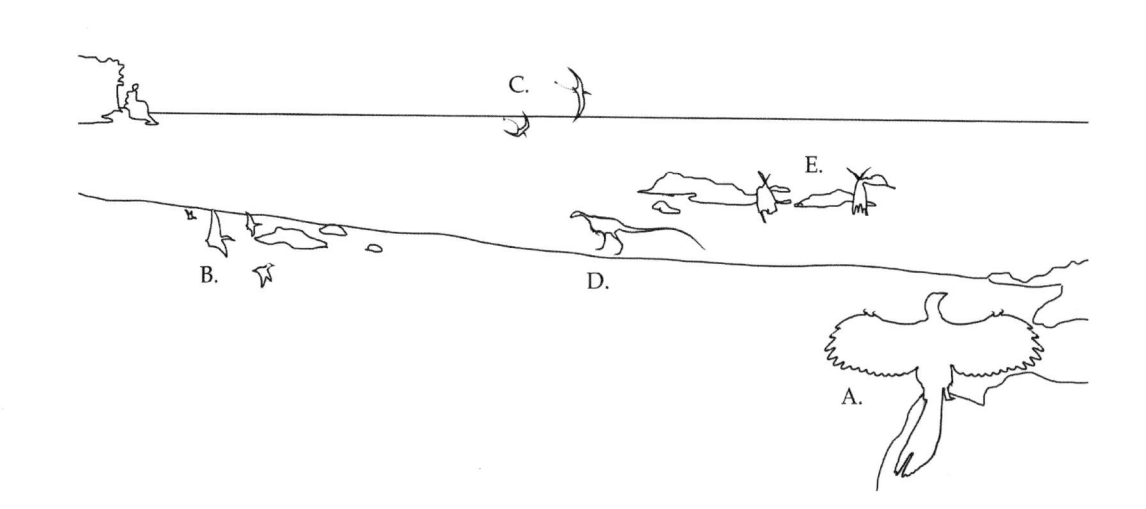

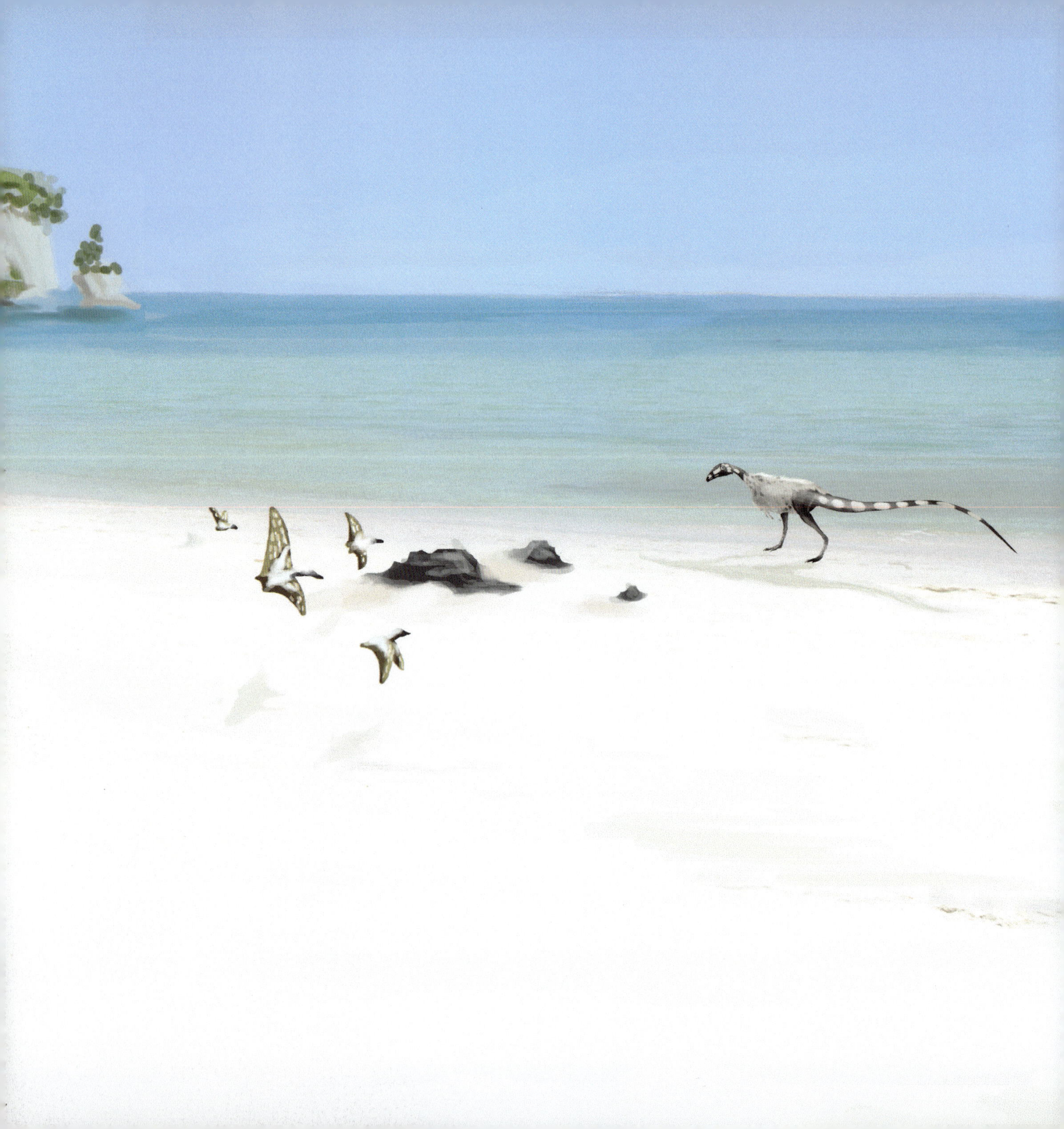

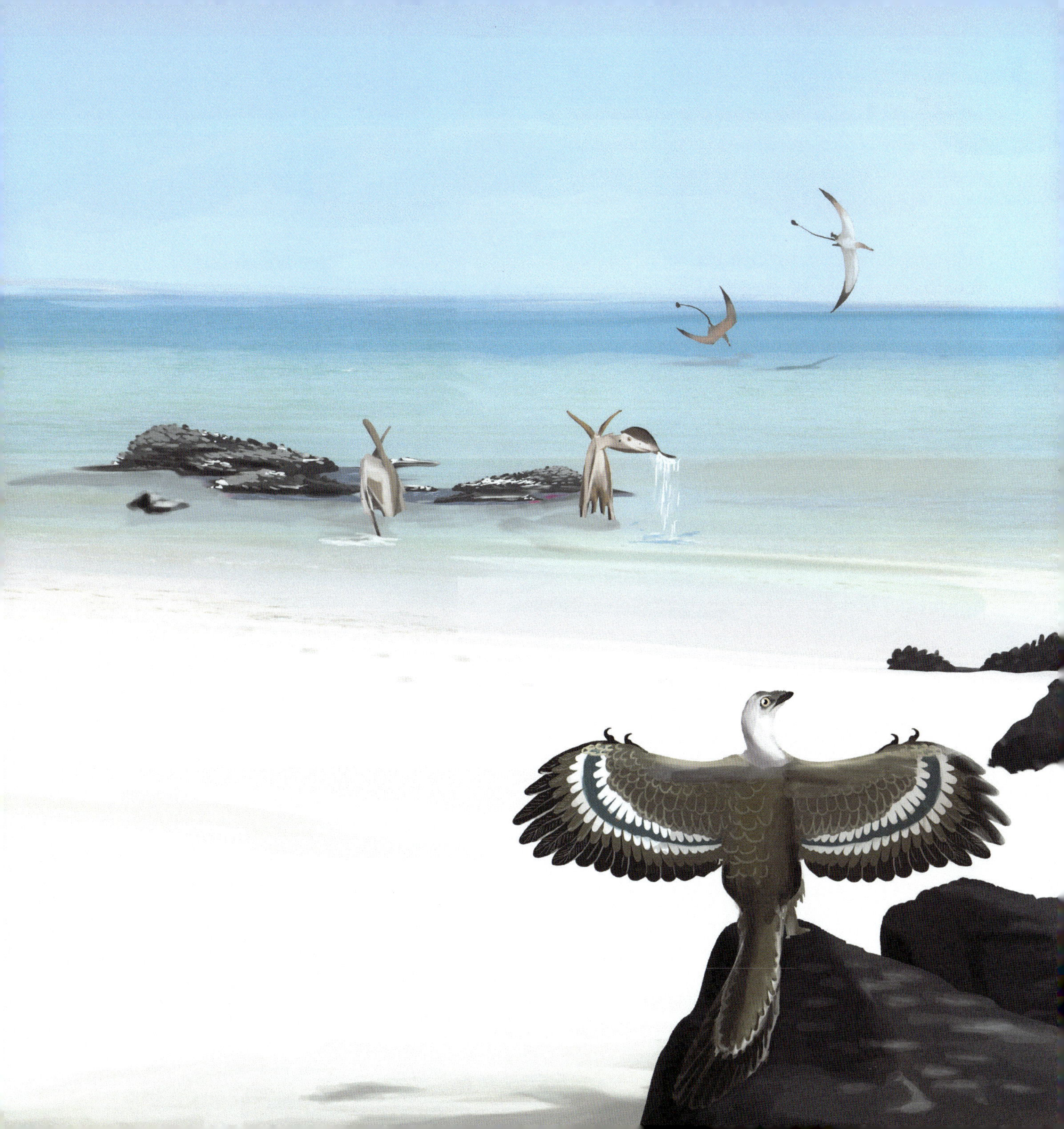

pterosaurs, has dramatically pared down the list of Solnhofen species. It has taken a wealth of comparative work and statistical studies, especially by scientists like Christopher Bennett (e.g. Bennett, 2013), to begin to understand which sets of juvenile specimens should be classified in which species, and to untangle the complicated list of synonymous names that have resulted from years of "splitting" juvenile fossils.

It may seem like this new era of research has resulted in a steep decline in overall stem-bird diversity in general, and especially pterosaur diversity in particular. In reality, we are really just re-thinking what diversity means in the context of their unique life histories. Diversity among stem-birds came mainly in the form of role shifting during their long period of slow growth, rather than the kind of high species diversity seen in modern birds and mammals, which have much faster growth rates. Stem-birds were highly successful to the point of dominating ecosystems on land for hundreds of millions of years, and they were able to do so in part thanks to the ability of each species to inhabit several different niches in the same ecosystem.

Fig. 1.

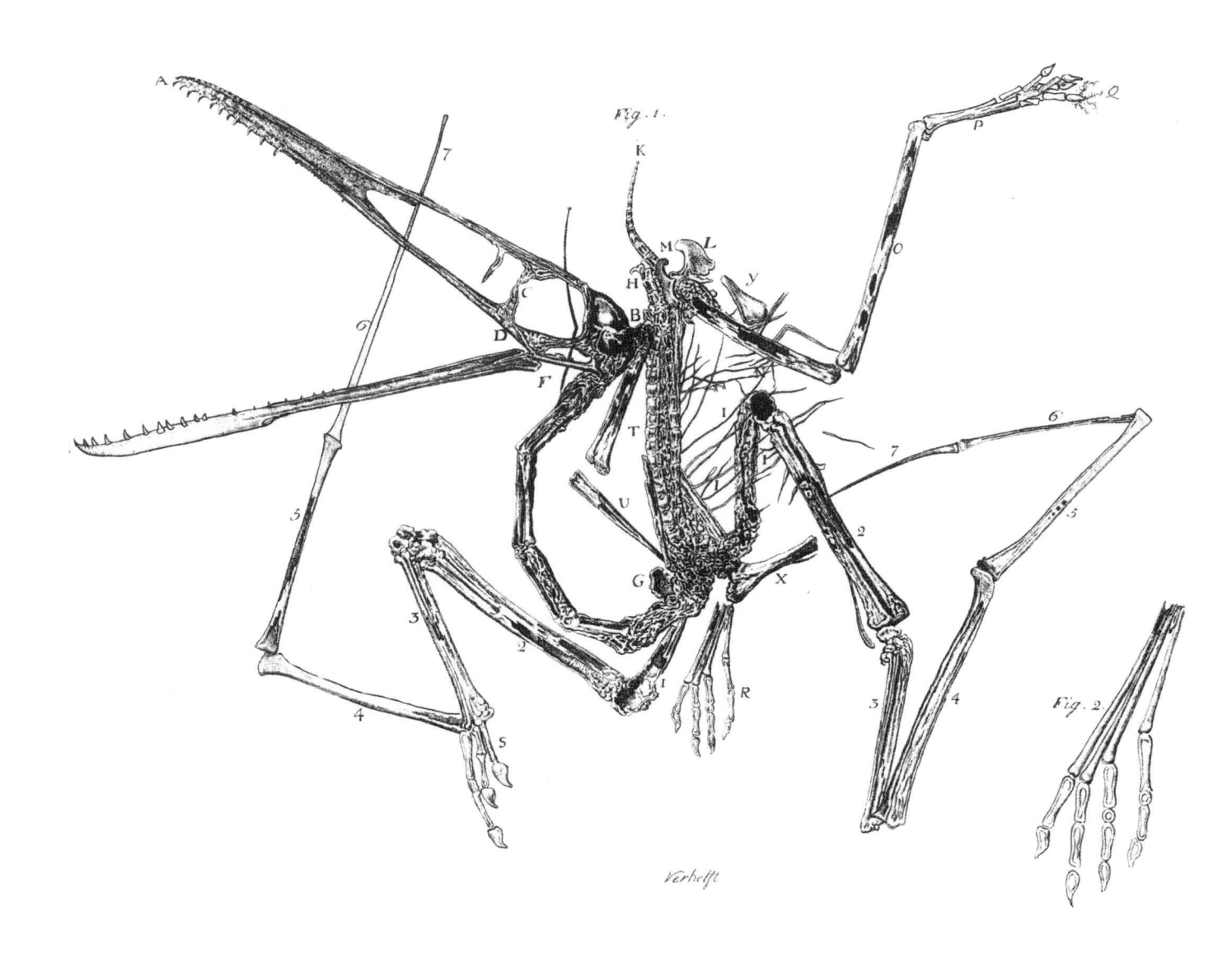

Fig. 2.

Verhelst

Pterosaurs

Traditionally referred to as "flying reptiles" (German "flugsaurier"), pterosaurs seem to have been among the first major groups to diverge from the bird lineage after it split from the ancestors of crocodilians. Pterosaurs are popularly known as pterodactyls, though technically this term has come to be restricted to either the genus *Pterodactylus*, or the clade of short-tailed pterosaurs, *Pterodactyloida*. The word

Ctenochasma elegans *in flight.*

"pterodactyl" (meaning "wing finger" after the part of the hand responsible for supporting the wing membrane) was coined by Cuvier for the very first known pterosaur specimens. Scientists in the 19th century continued to use it generally for all pterosaurs, long- or short-tailed, so the broader application of the term as a common name for all pterosaurs is not necessarily an error. As of this writing, the clade *Pterosauria* lacks a generally agreed-upon definition, but it may be useful in the future to define two separate groups: a clade *Pterosauria*, for the clade of stem-birds with a wing membrane used for flying or gliding, and a clade *Pterodactyla*, for all pterosaurs with an elongated fourth finger used to support the membrane.

The origins of this strange group are still largely a mystery, and there is no pterosaurian equivalent of *Archaeopteryx* to help identify what their ancestors were like. In fact, a few researchers maintain that pterosaurs were not stem-birds at all, but rather members of the archosaurian stem group, having evolved from ancestors more primitive than the common ancestor of crocodilians and birds (Hone & Benton, 2007). It is possible that many of the bird-like features present in the skeletons of pterosaurs can be explained by convergent evolution (and some features most certainly are convergent with birds and evolved independently due to the demands of

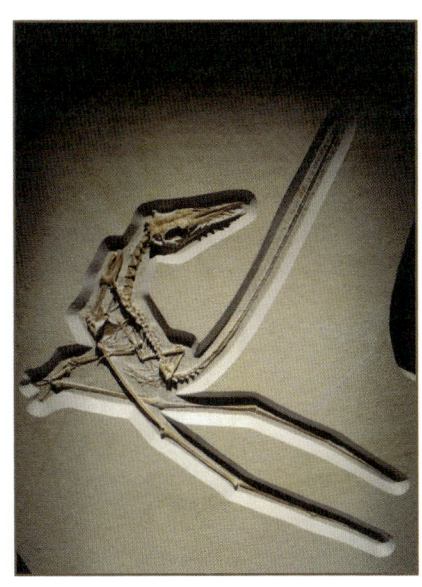

Previous page: Original copper engraving print of the Mannheim specimen by Egid Verhelst II and published by Cosimo Collini in 1784.

Fossil specimen of Rhamphorhynchus muensteri *found in the Wintershof Quarry near Eichstätt.*

Cast of the Mannheim specimen in the collections of the Carnegie Museum of Natural History, Pittsburgh.

active flight), but for now, the weight of the evidence and most thorough phylogenetic studies supports the idea that pterosaurs were stem-birds (Nesbitt, 2011), and shows that stem-birds produced successful flying lineages at least twice.

The taxonomic history of pterosaurs in the Solnhofen limestone is confused by the fact that young juvenile, or "flapling", specimens were once thought to represent unique species. Research by paleontologists like Peter Wellnhofer and Christopher Bennett has overturned this assumption and shown that, rather than a high diversity of small pterosaurs, the Solnhofen was populated by a small number of species, each occupying a wide variety of niches, and changing their ecological roles as they grew. Many pterosaur flaplings in the Solnhofen looked similar, except for minor differences in proportions, tooth count, and jaw shape, differences which became progressively more exaggerated as they grew into adults that differed vastly in shape and niche. In effect, flaplings of larger species took the place of small pterosaur species in the

Life restoration of Pterodactylus antiquus *based on the Mannheim specimen.*

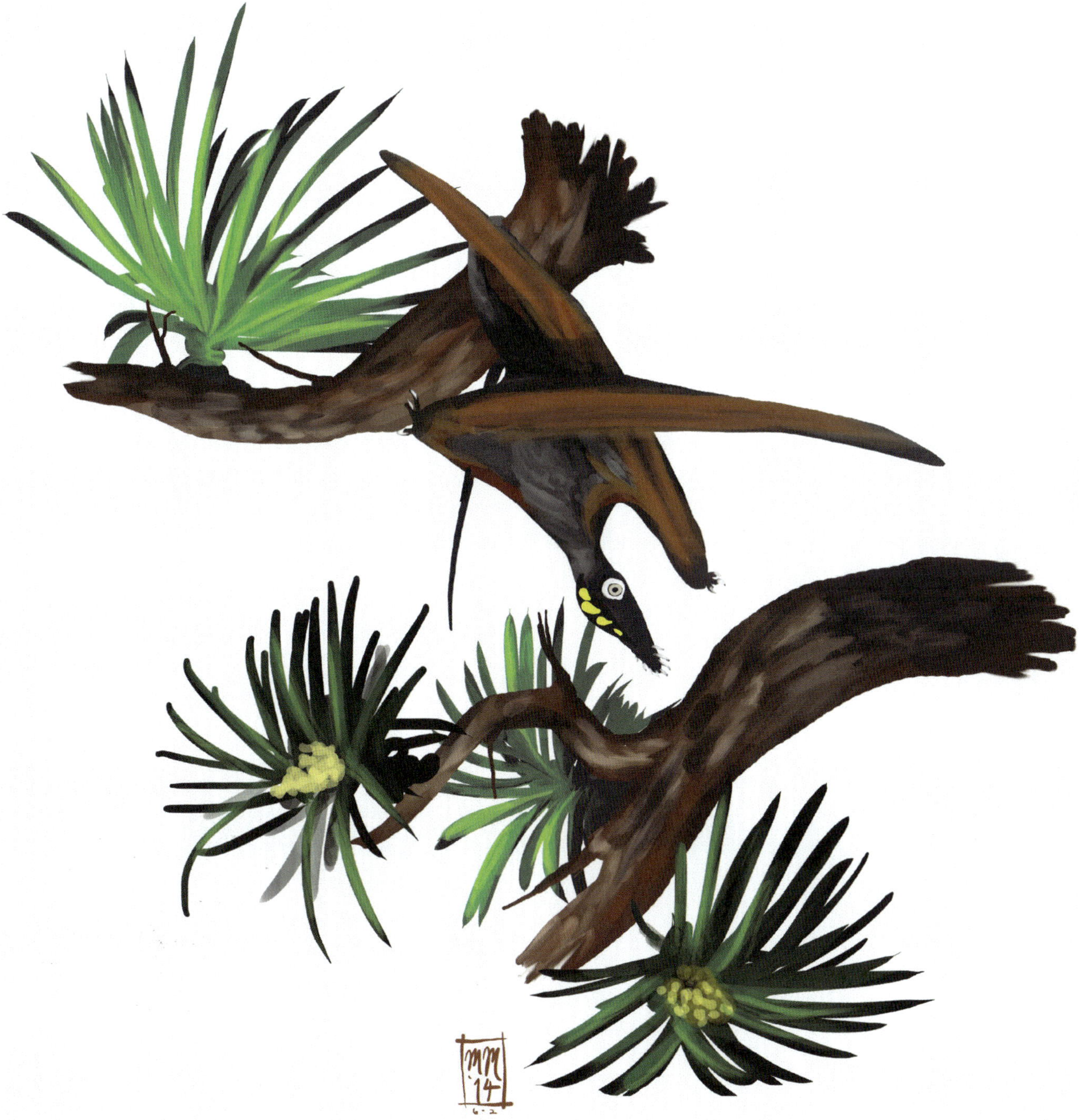

ecosystem (Unwin, 2006). Flaplings of all these various species probably acted together like a single "species" of small, generalist flyers, feeding on small arthropods like insects and crustaceans in and around the coastal waters. As the flaplings of each species grew and diverged from one another in lifestyle, so did their more obviously unique features, like jaw length and shape, tooth shape and orientation, and crest shape.

This variation across age groups means that the seashores of the Solnhofen archipelago would have been populated by a large number of generic-looking small baby pterosaurs flitting about in the surf and sand among their larger, more diverse, and more spectacular-looking parents.

"The Very Singular Skeleton"

The specimen named as the **holotype** (original) *Pterodactylus antiquus* was the first pterosaur fossil ever found. Its fossil remains came to light in the middle of a fascinating period of scientific growth in paleontology, as fossils began to be recognized as the remains of extinct organisms rather than interesting rock patterns or the bones of fantastic creatures that still inhabited the ocean depths.

Sometime during the 18th Century, the first *Pterodactylus* specimen was uncovered by quarrymen from the lithographic limestone near Eichstätt, today a part of Bavaria, but at the time a

Portrait of Cosimo Collini.

Portrait of Elector Charles Theodore by Anna Dorothea Therbusch, 1763.

semi-independent principality of the Holy Roman Empire. The fossil, a complete skeleton crushed flat, somehow came into the possession of the Count John Freiderich Ferdinand of neighboring Pappenheim. Ferdinand donated it to the nature cabinet of Prince-Elector Charles Theodore (1724-1799), located in Mannheim. It became known as the Mannheim specimen after that.

Nature cabinets and other "cabinets of curiosity" were the precursors of the modern concept of the natural history museum. These "cabinets" were actually rooms or series of rooms in which wealthy nobles with an interest in science could display and organize their private collections. Theodore's nature cabinet was curated by his secretary, a Florentine nobleman scholar named Cosimo Collini (1727-1806), who had previously served as secretary for, and was a good friend of, the famous Enlightenment thinker Voltaire. Collini left Voltaire's service after five years,* and Voltaire secured Collini his position with Elector Theodore in 1756.

*Collini had included a joke about Voltaire's niece and housekeeper, Madame Denis, in a letter accidentally left open on his desk. When she discovered the letter, she was offended by the joke, prompting Collini's resignation (Standish, 1821).

The exact date on which the Mannheim specimen entered Theodore's collection is unknown. In 1767, ten years after Collini began his tenure with Theodore, an inventory was taken of the cabinet, in which the specimen is not mentioned, so the Mannheim specimen must have been acquired between that date and 1784, when Collini published his description of it (Ősi et al., 2010).

The fossilized animal obviously struck Collini as significant, as he recognized that it was like nothing ever seen before. A bird-like head and neck, a small, mammal-like body and tail, long, reptilian claws and teeth and, most curiously, one incredibly long finger on each hand. Collini was at a loss, totally unable to identify even its class (at the time, the notion had not yet been widely popularized that fossil animals were new forms of life, rather than differently proportioned representatives of living animal types). Unable to sort out the fossil's chimerical nature and identify its known modern forms, Collini speculated that such bizarre animals must exist in the mysterious depths of the sea, and so have avoided scientific notice. He did note that in some ways it was bird-like, and in some ways appeared bat-like, but rightly concluded it was not a good match for either. For his paper, Collini enlisted Egid Verhelst II to create a copper engraving of the fossil as an illustration to accompany his paper. Verhelst thus became the first person to ever draw a pterosaur (Taquet & Padian, 2004).

Mannheim Palace in 2013, the original home of the first Pterodactylus antiquus *specimen.*

The French scientist Johann Hermann became aware of the fossil sometime after Collini published his description. In 1800, Hermann was concerned that, because Napoleon Bonaparte's occupying army was confiscating interesting and culturally significant artifacts, the fossil would likely be taken from the Mannheim Palace nature cabinet and sent to France. Hermann wrote a letter to the top French scientist of the day, expert anatomist Georges Cuvier, letting him know that the strangest fossil he'd ever seen, or as he put it a "very singular skeleton", was likely coming Cuvier's way. Hermann became the great-grandfather of all pterosaur-related paleoart by providing the first life illustration of the small creature. He guessed that the long fingers supported a wing membrane, and that the creature was a long-snouted, bat-like mammal. Furthermore, he pointed out that some stories coming out of China indicated that similar creatures might still be alive in the jungle there. Cuvier was able to obtain a copy of Collini's paper, and replied that the reports from China were nonsense, but was intrigued by the detailed diagram provided by Collini. Cuvier agreed that the finger probably formed a wing, though he disagreed with Hermann's classification, believing it to be a reptile rather than a mammal. Cuvier eagerly awaited the opportunity to study the actual specimen (Taquet & Padian, 2004), but his efforts were hampered when he tried to locate it.

Napoleon had not actually raided the nature cabinet. Elector Theodore had died in February of 1799, and in 1802, his entire natural history collection was brought to Munich, much to the dismay of Cosimo Collini, who felt a personal connection and obligation to its care. Once in Munich, the Baron Johann von Moll managed to convince the French to grant the Bavarian collections an exemption from being confiscated. By the time Cuvier learned of the fossil's whereabouts and wrote to von Moll asking to study it, von Moll replied that it was missing. This didn't deter Cuvier, who published his own description in 1809 based on the previous reports from Collini and Hermann. Without ever having seen the specimen, he gave it a name: *Petro-Dactyle*, literally "stone finger." In a later reprint, Cuvier corrected his apparent spelling error, and emended the name to *Ptero-Dactyle*, "wing finger" (Taquet & Padian, 2004).

Portrait of Georges Cuvier by François-André Vincent.

Interpreting the Pterodactyl

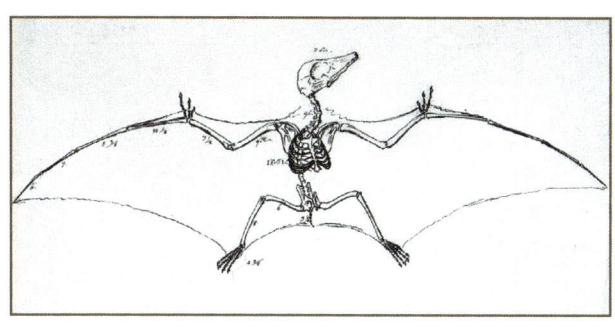

Sömmerring's 1817 restoration of P. brevirostris *as a proto-bat.*

Von Moll could not find the Mannheim Specimen in his collection because other scientists had already begun the process of studying and describing it. First was Johann Friedrich Blumenbach, who hypothesized that it was a type of shorebird in 1807. (Blumenbach is also notable as one of the first scientists to ever study a similarly perplexing chimera of an animal, the platypus). Next, German anatomist Samuel Thomas von Sömmerring studied the fossil in depth, concluding with a lecture delivered on December 27, 1810 that the specimen was a mammal. Shortly thereafter, he wrote to Cuvier apologizing for the situation, claiming that he had only just been told of Cuvier's request for information. Sömmerring told Cuvier that the latter's belief that the ptero-dactyle was a reptile was due to inaccuracies in Collini's description. Cuvier was frustrated by the lack of communication on the part of the German scientists and their reluctance to give him access to the specimen. Though Cuvier was sent a cast of the ptero-dactyle in 1818, he did not consider this sufficient for proper study (Taquet & Padian, 2004).

Portrait of Samuel von Sömmerring, Wellcome Library, London.

Sömmerring published the contents of his lecture in 1812, in a paper entitled "On an *Ornithocephalus*, or the unknown Beast of Antiquity." In this paper, he took the opportunity to give it its first formal species name, *Ornithocephalus antiquus*. *Ornithocephalus* means "bird head". Sömmering, believing it to be a kind of bat, gave it this name based on the feature he saw as primarily distinguishing it from other bats. Sömmering disagreed vehemently with Cuvier regarding the identity of the fossil, arguing that it was a mammalian link between birds and bats, not a reptile (though a link in "affinity" rather than the modern evolutionary concept of relatedness between animal types). Sömmering, however, notoriously misidentified several parts of pterosaur anatomy, especially in a restoration of a second specimen that he described in 1817 as a new species (now recognized as probably a flapling, probably belonging to *Ctenochasma elegans*).

By the mid 19th century, general consensus was reached in favor of Cuvier's view, that pterodactyls were reptiles. However, several other, sometimes idiosyncratic, hypotheses were proposed by various scientists during this period. In 1830, the herpetologist Johann Georg Wagler re-interpreted pterodactyls not as flying reptiles or primitive bats, but as amphibious reptiles. He envisioned the long, spindly bones of the foot fanned out and supporting webbed feet or paddles, and the long wing finger forming the center of a long flipper. It seems that other scientists did not take Wagler's suggestion seriously, and no others took up his hypothesis. Further confusion was caused by the supposed discovery that pterodactyls had hair or feathers. A downy coat was first noticed surrounding fossils of *Pterodactylus crassirostris* (=*Scaphognathus*) and *P. medius* (=*Germanodactylus*?) in 1831 by the paleontologist Georg August Goldfuss. On close examination of the limestone slabs of these pterosaur specimens, Goldfuss believed that he could identify small pits in the area of the wing membrane, which he interpreted as follicles, and tiny striations, interpreted as hair and/or feathers, including longer filaments forming a sort of "mane" on the neck. This would obviously have been taken as evidence against Cuvier's reptile

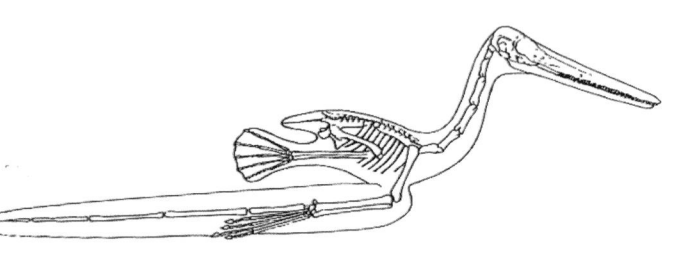

Wagler's 1830 restoration of P. antiquus as an amphibious reptile.

Newman's 1843 restoration of Scaphognathus crassirostris *(top),* Ornithocephalus brevirostris *(bottom), and* Pterodactylus antiquus *(background) as marsupial "bats".*

Early restoration of "furry" but reptilian pterosaurs. Clockwise from left: P. spectabilis (=P. antiquus), R. muensteri, S. crassirostris. *By Charles Whymper for Knipe's "Nebula to Man", 1905.*

hypothesis, and other scientists attempted to explain this discrepancy, as well as the apparent absence of scales. The Swiss geologist Louis Agassiz stated his opinion that these striations in the rock matrix were the details of wrinkles in the wing membrane. English paleontologist William Buckland agreed, and in his *Bridgewater Treatise*, wrote that scales would have hindered pterodactyls in flight, and that they appear to have had naked skin. In 1843, British naturalist Edward Newman disagreed with these opinions, and in his own words, dared to contradict the great Cuvier. Newman doubted the identification of the striations as wrinkled skin, and scoffed at the idea that an animal lacking scales should be classified as a lizard. Also unlike lizards, pterodactyls appeared to have only four fingers. This, combined with the presence of a skeletal element Newman identified as a marsupial bone, led him to re-interpret pterodactyls not as bats or reptiles, but as furry, flying marsupials (Newman, 1843).

Despite these challenges, Cuvier's interpretation eventually won out, especially after finding prominent supporters in the English speaking world like Buckland and Sir Richard Owen. Owen consulted on the famous Crystal Palace sculptures of prehistoric animals, where four sculptures were produced of "pterodactyls of the Oolite" perched on a cliff. These were obviously reptilian, wyvern-like creatures with prominent lizard-like overlapping scales (contrary to the evidence from fossils as pointed out by Buckland), robust hind legs, and huge, billowing, sail-like wing membranes.

The debate, though heated (and possibly because it was so heated), between Cuvier, Sömmering, and other early anatomists was a watershed moment for biological science. In order to support their points, these anatomists for the first time were not using general form and ecology to try and classify a new animal; rather, they were pointing out specific features or suites of features, and using rigorous anatomical comparison to ally the specimen with one group or another. Around the pterodactyl was born modern comparative anatomy and a classification system based on shared characteristics, laying the groundwork for classification based on evolutionary and genetic relationships rather than general form (Taquet & Padian, 2004).

In the following decades, many more pterosaur specimens were found, most of them classified as species of *Pterodactylus* (even gigantic British forms like *Ornithocheirus*). Today, these are recognized as a diverse array of distinct species. The presence of feather-like filaments was confirmed in *Rhamphorhynchus* by Wanderer in 1908 and later in *Pterodactylus* by Broili in 1938, but their exact nature was still controversial and all but ignored in popular depictions of pterosaurs, such as those of artist Zdeněk Burian. It was not until the discovery of an incontrovertibly downy rhamphorhynchid, *Sordes pilosus*, from the Soviet Union in 1971 did the idea of "hair" on some pterosaurs gain consensus. In 2002, examination of the original *Scaphognathus* specimen using ultra-violet photography confirmed that Goldfuss' interpretations of "hairs" and other soft tissue preservation had been right all along (Tischlinger, 2002).

Major discoveries relating to *Pterodactylus antiquus* itself did not really resume until the late 1990s, having given way to research into newer, bigger, and more spectacular pterosaurian forms. In 1996, Christopher Bennett published a paper on an excellent specimen with extensive soft tissue impressions, including an outline of the body. This confirmed not only that *P. antiquus* possessed a uropatagium (wing membrane between the legs and tail), but that it extended onto the foot and between the toes, creating webbed feet, and then up the other side in a narrow strip probably connecting to the wing itself. There is a long and ongoing debate about wing extent and attachment point in pterosaurs: narrow-chord with attachment high on the leg, broad-chord with attachment at the ankle, or even an extremely bird-like arrangement with the wing attached to the hip, and the legs tucked up under the body in flight, as suggested first by Kevin Padian in 1983.

In *Pterodactylus*, at least, it seems to have been a bit of both: the wing itself was narrow, extending probably not far past the elbow, but then curving backward to form a narrower section along the leg and finally anchoring on the foot. This same specimen also preserved a dense covering of bristly filaments along the neck, down-like structures now known as pycnofibres. While often considered a form of unique hair-like filament, the discovery that most groups of stem-birds had various kinds of filaments covering parts of their bodies makes it more likely that pycnofibres were a type of primitive feather related to the more complex flight feathers of modern birds.

Another specimen described in 2003 and studied extensively under ultra-violet light confirmed these new discoveries about the animal's life appearance. German paleontologist Peter Wellnhofer had noticed in the 1970s, a small lobe extending from

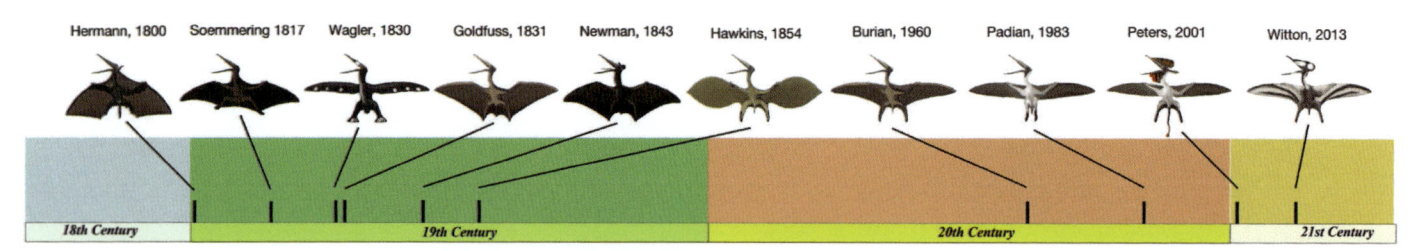

Timeline of notable life restorations of Pterodactylus.

the back of a specimen's skull, and suggested *Pterodactylus antiquus* may have had a fleshy wattle-like crest composed entirely of non-bony tissue. Bennett, in 1996, doubted this finding, suggesting it was a preservation artifact, but the 2003 study confirmed that *P. antiquus* had a backward-pointing wattle, made up of twisted fibrous tissue (Frey et al., 2003). Bennett (2013) suggested this "occipital cone" would better be termed a lappet. However, in birds, the term "lappet" is already used for certain types of broad, folded skin caruncles. "Wattle" might be a more appropriate term, or simply the more general "caruncle".

More than perhaps any other stem-bird, *Pterodactylus antiquus* is connected to the history of paleontology, of paleoart (at least in terms of how ancient life was interpreted and restored), the history of museums, and with many prominent historical figures and events. And it's still yielding secrets today, clues that are helping us understand key aspects of pterosaur biology and life appearance.

Pterodactylus antiquus

"Ancient Pterodactyl"

Classification:
  Class: Stem-Aves
  Order: Pterosauria
  Suborder: Pterodactyloidea
  Family: Pterodactylidae
  Genus: *Pterodactylus*
  Species: *Pterodactylus antiquus*

Formations: Solnhofen, Moernsheim

Age: Early - middle Tithonian, 151.5 - 150 million years ago

Wingspan: Up to 1 meter

Main diet: Small fish

Over 30 specimens of *P. antiquus* have been identified which are still referred to this genus. Many others were once assigned to either *P. antiquus* or *Pterodactylus* in general, but are now considered different species or genera. Even *P. kochi*, long thought to be a *Pterodactylus* distinct from *P. antiquus*, has been shown to be a growth stage of this species.

Having so many well-preserved specimens to work with has allowed researchers to reliably reconstruct not just the detailed skeletal anatomy, but soft tissue anatomy as well. *P. antiquus* differed from similar species such as *Germanodactylus rhamphastinus* and *Ardeadactylus longicollum* in the possession of a unique suite of skeletal traits, especially concerning the shape, position, orientation, and number of teeth.

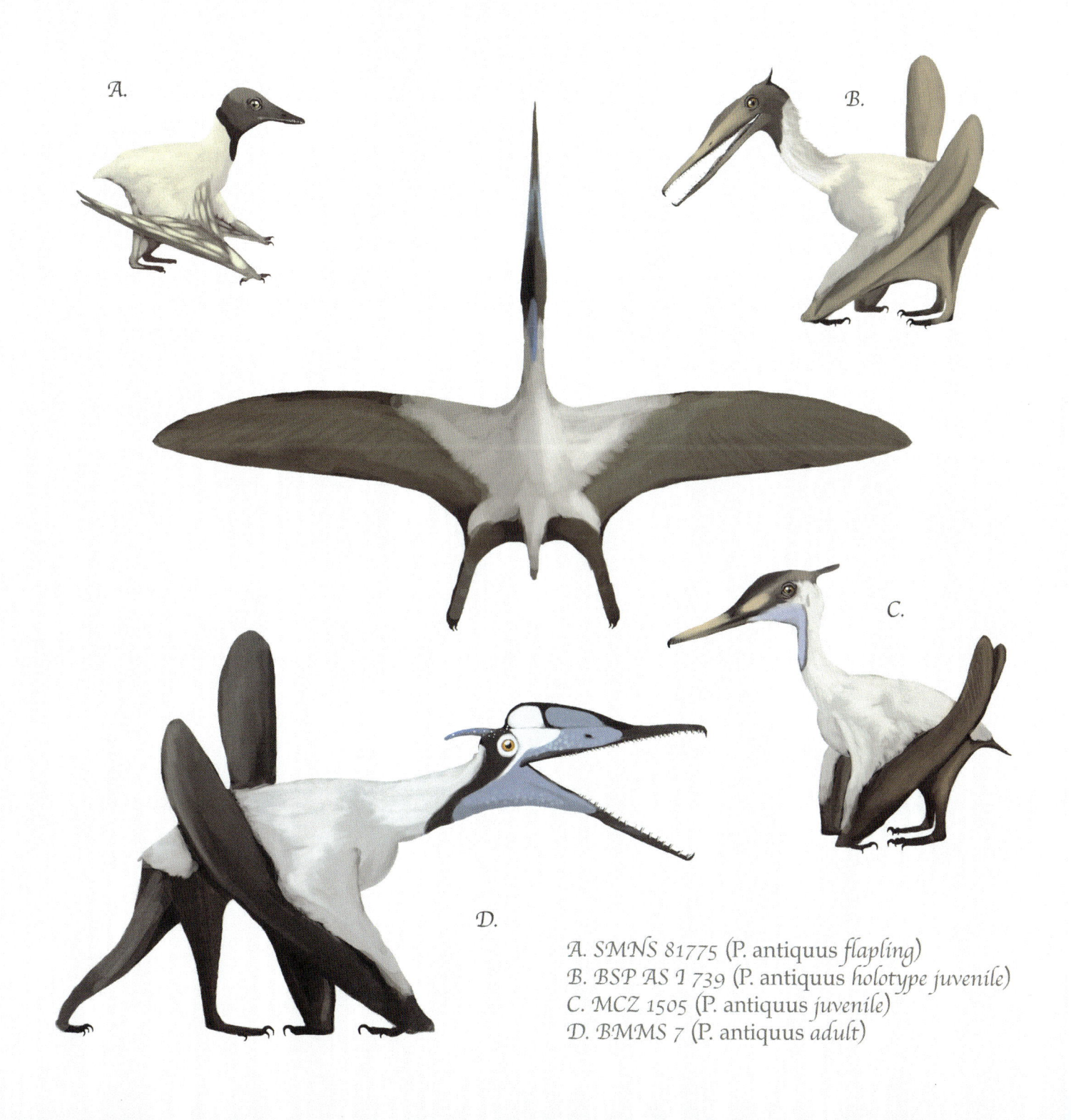

A. SMNS 81775 (P. antiquus *flapling*)
B. BSP AS I 739 (P. antiquus *holotype juvenile*)
C. MCZ 1505 (P. antiquus *juvenile*)
D. BMMS 7 (P. antiquus *adult*)

The skull of *P. antiquus* was long and narrow, mainly straight but with a very slight upward curve. The snout (often incorrectly referred to as a "beak") was long and thin, with about 90 teeth present in adults. The teeth occupied the front two thirds of the upper jaw, and were present behind the position of the nostril. In adults, the front teeth were much larger than the tiny back teeth, but in flaplings this size difference was less drastic. A hooked beak, about the length of a tooth, capped the extreme tip of the upper jaw. A low midline crest began to grow from the skull when the animal reached about half a meter in wingspan, and became larger with age. The bony core of the crest lacked striations as seen in some other species, which may indicate the total size of the crest was generally smaller, with a less extensive keratin extension. A long, thin caruncle (apparently single, not paired) extended from the back of the head. This caruncle (also variously called an occipital cone, chignon or lappet) was apparently made of a base of twisted fibers bound together by soft tissue. (Bennett, 2013).

The neck was longer than many related species, due to elongated vertebrae. A small gular sac (throat pouch) was present, but was not large enough to trap fish as pelicans do. The throat pouch may have been used to transport small amounts of food, or for display. Overall, the most similar contemporary species is *Germanodactylus rhamphastinus*, which shares many of these features, but differs in some details. (Bennett, 2013). Like *Scaphognathus*, *Pterodactylus antiquus* seems to have had relatively long feathers on the neck compared with the rest of the body (Frey et al., 2003). Originally interpreted as a "mane", this probably served instead like an aerodynamic shell, streamlining the neck smoothly into the body, and making it look shorter than it would otherwise.

The shape of the jaws and teeth suggest *P. antiquus* fed mainly on small fish. It may have flown low over the water and dipped into schools of fish, and/or fished while floating on the surface of the water (Bennett, 2013). Like other pterosaurs, *P. antiquus* grew slowly, and smaller individuals probably fed differently than larger ones. The differences between juveniles and adults, mainly in the shape of the snout and arrangement of teeth, support this view. Flaplings had relatively short but still rather narrow snouts, with proportionately larger teeth throughout the length of the jaw. Juveniles may have been more generalist beachcombers, picking and probing in the sand and surf, while the larger individuals flew or swam after fish.

A Note on Nomenclature: Cuvier named *Petro-Dactyle* without a type species. For this reason, some have claimed that *Ornithocephalus* is technically the correct genus name for this animal (Martill, 2014). However, the ICZN does not actually require a type species to be fixed at the same time as the genus name is coined for names created before 1931. The ICZN does mandate simply removing marks like hyphens, but not a mandatory Latinization of names (Article 32), making the later emendation of the name, *Ptero-dactyle* Cuvier 1809 to *Pterodactylus* by Rafinesque 1815, unjustified. The change from "*Petrodactyle*" to "*Pterodactyle*" does appear to have been a justified spelling correction, as Cuvier himself changed the spelling for later printings of his paper. The correctly emended name would therefore be *Pterodactyle* Cuvier 1809. Luckily, Article 33.2.3.1 of the International Code of Zoologican Nomenclature states that even if an emendation is unjustified, if it is in prevailing use and attributed to the original author, it is to be considered justified. This leaves the question of whether or not *Ptero-Dactyle* was intended as a formal scientific name in the first place. It is not clear that the name was initially intended to be an official scientific name, rather than a common or vernacular name, or non-Linnaean descriptive term. It seems that at least some contemporary scientists interpreted the name as being vernacular, such as Fischer 1813, who acknowledged Cuvier's term *Pterodactyle*, but gave it his own name, *Pterotherium*. However, Cuvier does seem to have intended the name as a label for a genus of animal. He used as the title of his paper, "Sur le squeletle fossile d'un *Reptile Volant* des environs Aichstedt, que quelques naturalistes ont pris pour un oiseau y et dont nous formons un genre de *Sauriens* sous le nom de *Petro-Dactyle*." (Cuvier, 1809), which translates as "On the fossil skeleton of a *Flying Reptile* from Eichstätt that some naturalists have taken for a bird, but which we interpret as a kind of *Saurian* under the name *Petro-Dactyle*." In an age when new names were often created without a formal binomen and in footnotes, this is "good enough."

Gnathosaurus subulatus

One of the most spectacularly bizarre of the pterodactyls was also, possibly, the first to actually be discovered. A specimen of a flapling pterosaur which is probably an example of *Gnathosaurus subulatus* was first found sometime between 1757 and 1779, overlapping with the potential discovery date of the Mannheim *Pterodactylus*. After being excavated from one of the Solnhofen quarries, the so-called Pester Exemplar ("Pest Specimen") came into the geological collection of Archduchess Maria Anna of Austria. Maria Anna was the daughter of the Holy Roman Emperor Francis I, and Maria Theresa, Queen of Hungary and Bohemia. Despite her high birth, Maria Anna suffered from a physical disability and poor health throughout her life. She began collecting fossils and other geological specimens in 1757 while recuperating from a bout of illness, and so must have acquired the Pester Exemplar at some point after that.

The unusual specimen was fragmentary and disarticulated, but was interesting enough to warrant a published description by Ignaz von Born, a leading scientist of the Holy Roman Empire and Maria

---

"POINTED JAW SAURIAN"

Classification:
  Class: Stem-Aves
  Order: Pterosauria
  Suborder: Pterodactyloidea
  Family: Ctenochasmatidae
  Genus: *Gnathosaurus*
  Species: *Gnathosaurus subulatus*

Formation: Solnhofen

Age: Early Tithonian, 151.5 - 150.8 million years ago

Wingspan: Up to 1 meter

Main diet: Small invertebrates (by spooning & straining)

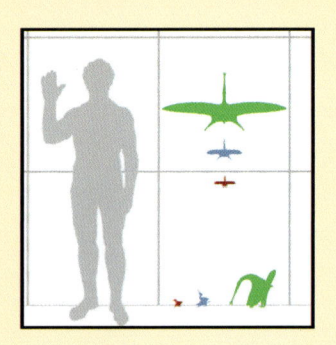

A. B St 1964 XXIII 100 (*G. subulatus? flapling*)
B. CM 11426 (*G. subulatus? juvenile*)
C. PTHE 1951 84 (*G. subulatus adult*)

Anna's mentor. Born, in his 1779 paper, concluded that the long, spindly appendages visible in the fossil (now known to be parts of the long wing fingers) were the legs of a crustacean. The specimen was not studied again until 1856, when German scientist Christian Erich Hermann von Meyer recognized it as a pterosaur. Meyer erected a new species of small pterodactyl, *Pterodactylus micronyx*, for a collection of specimens now known to be flaplings of various species, but of which the Pester Exemplar was the holotype.

Maria Anna's collection was sold in 1871 to the Royal University of Pest (today Eötvös Loránd University). When Peter Wellnhofer attempted to study the Pester Exemplar for his work on Solnhofen pterosaur specimens, he was told by curators of the University's collection that the specimen had been lost, and designated a neotype for *P. micronyx*. However, in 1982, the collection was re-organized, and during this process the original Pester Exemplar turned up, allowing the return of holotype status to possibly the earliest known pterosaur fossil.

During the 1990s and 2000s, S. Christopher Bennett attempted to systematically analyze the proportions of various flapling specimens from Solnhofen and match them with adults of the same species. In a 2013 paper, he concluded that *P. micronyx* specimens, including the Pester Exemplar, best matched the proportions of the recently named *Aurorazhdarcho primordius*, and he synonymized the two, using the new combined name of *Aurorazhdarcho micronyx*. Bennett noted that it was likely that this was the same species as *Gnathosaurus subulatus*, based on skull features of some former *P. micronyx* specimens. Though a rigorous comparison can not be conducted until either an adult skeleton of *Gnathosaurus* is found, or an adult skull of "*Aurorazhdarcho*" is found, "*A.*" *micronyx* remains the best match of all the Solnhofen flaplings for the skull and tooth anatomy of *Gnathosaurus subulatus*, so they are tentatively considered synonyms here.

Fossil jaws belonging to adult *G. subulatus* were first found in 1832. Long and narrow, they contained a tip full of thin, needle-like teeth forming a spoon-shaped rosette. The long snout with thin, pointed teeth caused scientists to consider this the fossilized jaw of a crocodile, and so it was named *Croco-

dilus "multidens" (the "many-toothed crocodile"). This name was later changed when Meyer transferred it to a new crocodilian genus, *Gnathosaurus*. Meyer regarded it as a prehistoric crocodilian similar to gharials and the stem-crocodilian teleosaurids. Once specimens of the similar *Ctenochasma* were found that preserved the skull and skeleton, it was realized that both were pterosaurs rather than crocodilians.

Like *Ctenochasma*, *Gnathosaurus* probably used their long teeth to sift through sand and trap or net small crustacean prey, though they lacked the tightly-packed teeth and unique palates of the more advanced ctenochasmatid *Pterodaustro* that allowed true filter-feeding.

A Note on Nomenclature: *Gnathosaurus subulatus*, originally interpreted as a crocodile fossil, was initially given the name *Crocodilus* "multidens". This means that, by today's standards, the well-known name *G. subulatus* would need to be changed to the older synonym, *G.* "multidens". However, Wellnhofer (1970) noted that Munster's original description gave very few details and could not be considered sufficient to create the name. Therefore, *C.* "multidens" is a nomen nudum.

Ctenochasma elegans

Specimen of a juvenile C. elegans *(AMNH 5147) in the collections of the American Museum of Natural History.*

The second pterosaur to be named, and the third specimen ever recognized, was the holotype of *Ornithocephalus brevirostris*. Another flapling specimen with an uncertain relationship to adults, this specimen made its way through various nature cabinets and Bavarian collections before Blumenbach alerted von Sömmerring to its existence. Sömmerring published a description of it in 1817, comparing it with the Munich specimen of *Pterodactylus antiquus*.

The *O. brevirostris* holotype (listed as nr. 29 in Wellnhofer, 1970) appears to correspond with the adult specimen later named *Ctenochasma gracile* (this was noted by Bennett in his 2013 revision of *Pterodactylus antiquus* without further comment). *Ctenochasma gracile* had previously been renamed *Ctenochasma elegans*

"Elegant Comb Gape"

Classification:
　Class: Stem-Aves
　Order: Pterosauria
　Suborder: Pterodactyloidea
　Family: Ctenochasmatidae
　Genus: *Ctenochasma*
　Species: *Ctenochasma elegans*

Formations: Solnhofen

Age: Early Tithonian, 151.5 - 150.8 million years ago

Wingspan: Up to 1.2 meters

Main diet: Small invertebrates (by straining)

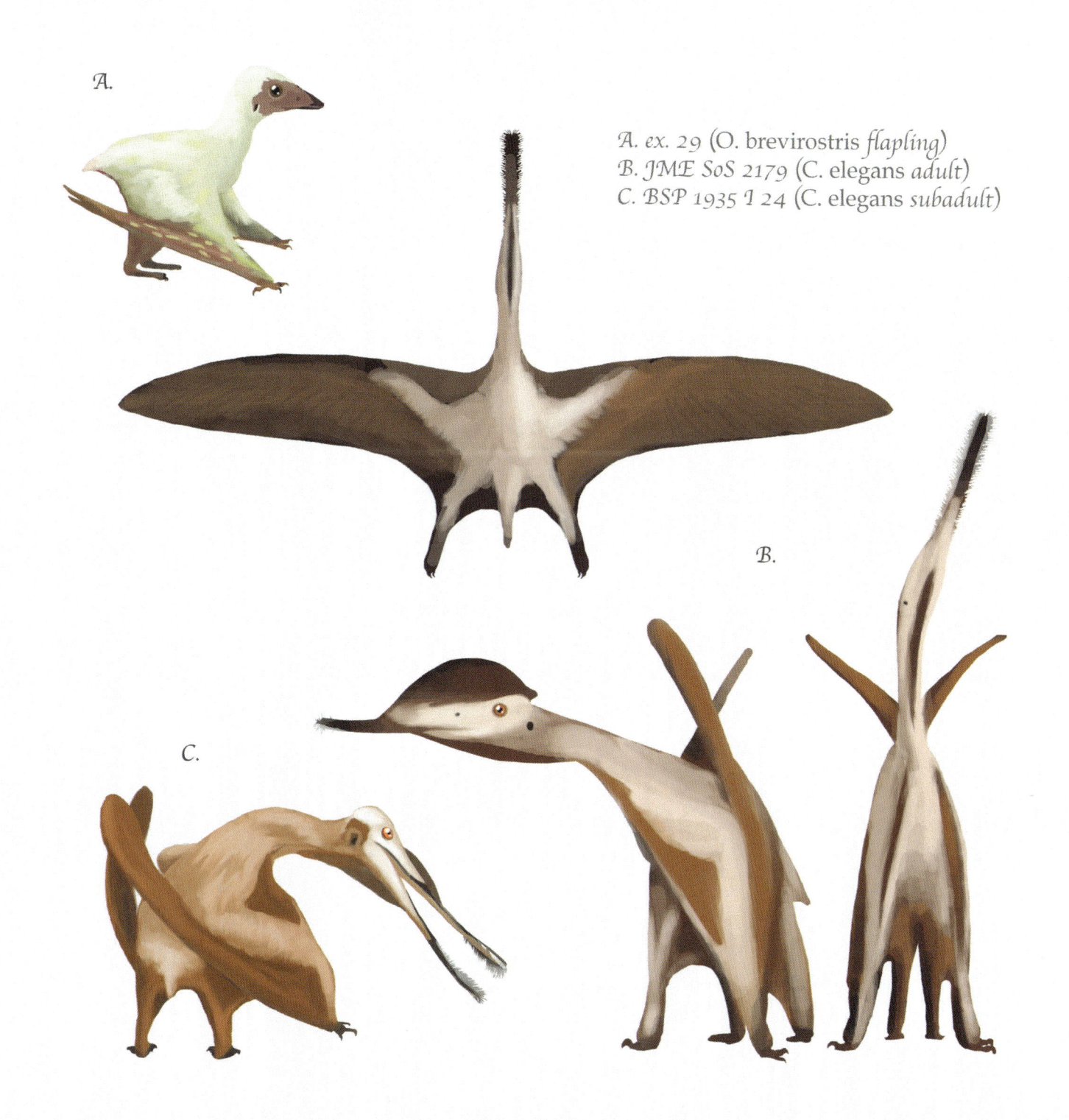

A. ex. 29 (O. brevirostris *flapling*)
B. JME SoS 2179 (C. elegans *adult*)
C. BSP 1935 I 24 (C. elegans *subadult*)

when studies suggested it corresponded with flaplings named *Pterodactylus elegans*, so it would be no great disruption of nomenclature to change this once more to the second-oldest pterosaur species name and recombine it as *Ctenochasma brevirostris*. This new name would be ironic, however, because adult *Ctenochasma* are the opposite of short-snouted!

Like *Gnathosaurus*, *Ctenochasma* had extremely long, slender jaws with numerous sideways-pointing, needle-like teeth. Also like *Gnathosaurus*, adult specimens were initially confused for crocodilians until the discovery of associated skulls and skeletons from subadult individuals. Oppel (1862) was the first to suggest that features of the partial jaws known at the time more closely resembled pterodactyls than crocodiles, but clear evidence that he was right was not found until 1924, when the first complete skeleton was uncovered (Bennett, 2007). *Ctenochasma* is notable for having its long, straining teeth present over more of the jaw length (the front half), with longer, more numerous teeth (over 400 in large adults). The jaws themselves had an upward curve, possibly to help scoop through sediment or schools of prey. In top view they were nearly rectangular; rather than tapering to a point, the jaw tips were broad and rounded (Bennett, 2007). The hind limbs were very long, and this, combined with long, broad feet, indicate that these pterosaurs may have spent time wading (and possibly occasionally floating) in shallow water as they foraged for prey. Adults bore a bony crest along the midline of the skull like many other pterosaurs, and this probably supported an even larger soft tissue crest in life.

Rhamphorhynchus muensteri

The first few pterosaur fossils known to science were all very small, owing to the fact that they came from young juveniles or subadult animals. So, when the fragmentary remains of a specimen that would have had a nearly two meter wingspan in life were discovered, it's no surprise that they were assigned a new species name: *Ornithocephalus giganteus*. It took many more specimens representing different parts of the body, and many more scientific names, before the true nature of this animal was discovered.

A second "giant" pterosaur specimen from Solnhofen, named *Pterodactylus grandis* by Cuvier in 1824, is occasionally listed as a synonym of *Rhamphorhynchus*. This name has been considered valid by several 20th century authors (e.g. Harris & Carpenter, 1996; Meyer & Hunt, 1999). If it proves to be the same species as *R. muensteri*, it would not be a candidate for nomen oblitum status. However, no serious study has yet been done to try and determine which, if any, well-known Solnhofen pterosaur species

"MÜNSTER'S BEAK SNOUT"

Classification:
  Class: Stem-Aves
  Order: Pterosauria
  Suborder: Rhamphorhynchoidea
  Family: Rhamphorhynchidae
  Genus: *Rhamphorhynchus*
  Species: *Rhamphorhynchus muensteri*

Formations: Nusplingen, Solnhofen, Moernsheim

Age: Late Kimmeridgian - mid Tithonian, 152.5 - 150 million years ago

Wingspan: Up to 2 meters

Main diet: Fish (by dipping?)

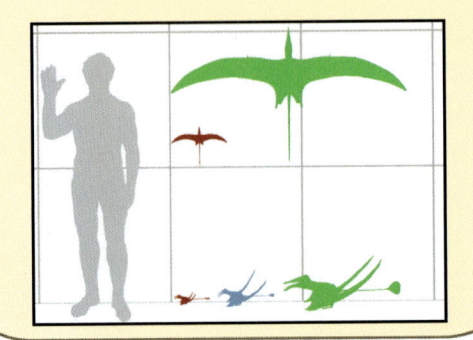

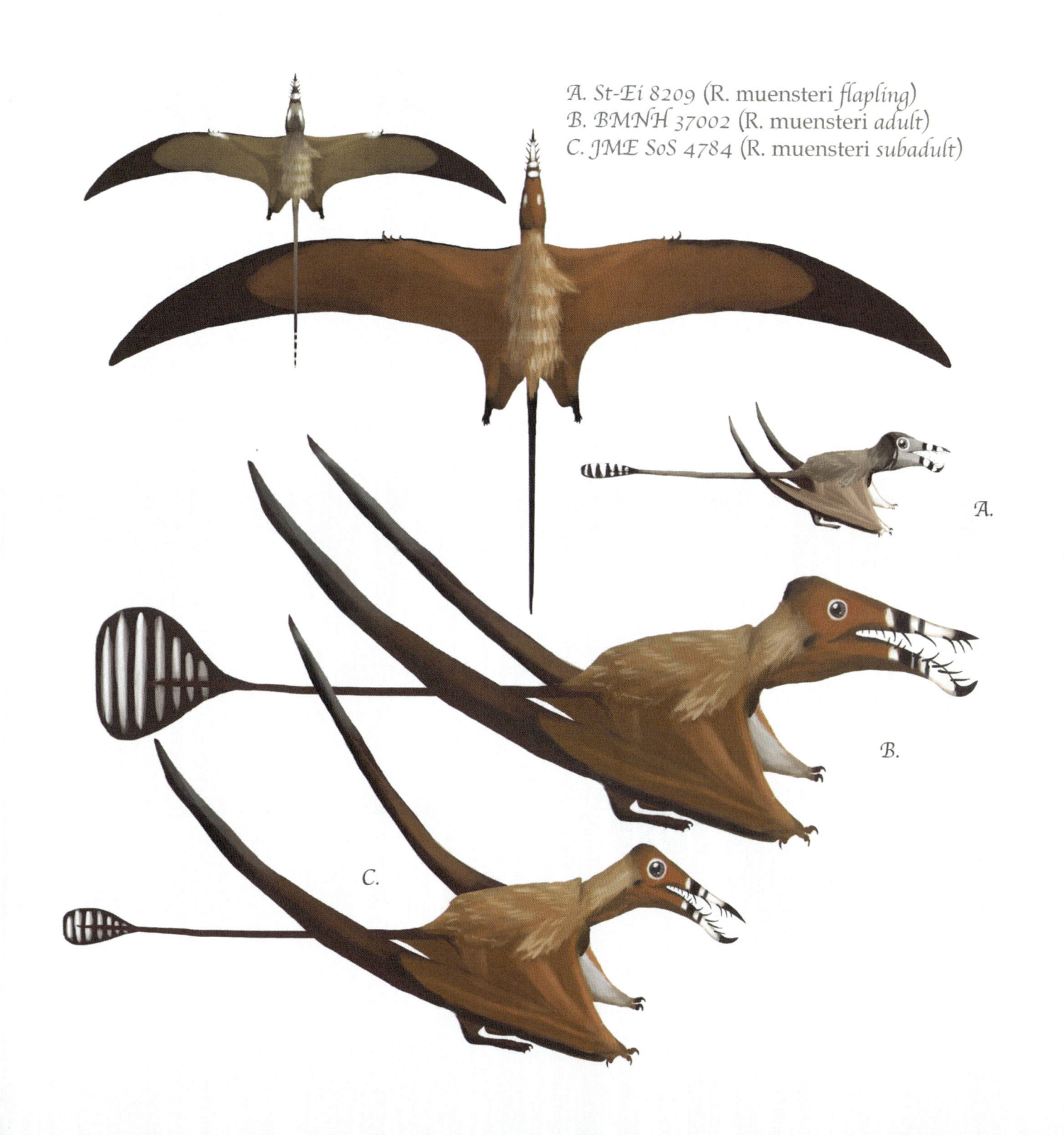

A. St-Ei 8209 (R. muensteri *flapling*)
B. BMNH 37002 (R. muensteri *adult*)
C. JME SoS 4784 (R. muensteri *subadult*)

this fragmentary specimen belongs to.

In 1831, Goldfuss described a new pterosaur specimen represented by a disarticulated skull and partial skeleton. Though preserved in top view, the skull was unlike other known pterodactyls in being fairly broad, with a relatively short but sharply tapering snout ending in a toothless portion that likely bore a beak. The tooth sockets protruded slightly to the side, and would have accommodated the long, curved teeth scattered around the skull. Despite these fairly significant differences in form from other pterodactyls, Goldfuss did not assign this specimen to a new genus, but merely a new species of "*Ornithocephalus*", as *O. muensteri* (for fellow paleontologist Georg zu Münster). A few years later, Münster himself described a small pterodactyl specimen with a long tail, the first clearly recognized with such a feature, as *Pterodactylus longicaudus*. In 1846, Meyer collected all of these beak-snouted, snaggletoothed, long-tailed pterosaurs and placed them into their own genus separate from *Pterodactylus/Ornithocephalus*: *Rhamphorhynchus*, with Münster's *R. longicaudus* as the type specimen.

The Edinburgh Specimen of R. muensteri, *held at the National Museum of Scotland.*

Subsequent fossil finds showed that *Rhamphorhynchus* were indeed the largest known Solnhofen stem-birds, with wingspans of over two meters in some large adults, making it likely that the early, fragmentary specimens like *O. giganteus* belonged to it. Like *Ptero-*

Specimen of R. muensteri *in the collection of the Royal Ontario Museum.*

dactylus, the genus *Rhamphorhynchus* once included a wide array of species which differed mainly in size and proportion, but later studies showed that these were better interpreted as growth stages or year classes of a single species, the oldest available name for which is *R. muensteri* (not counting dubious species names based on poor specimens) (Bennett, 1995).

R. muensteri is also among the most common Solnhofen pterosaurs, with over 100 specimens having been found, many of them complete. *R. muensteri* of all sizes seem to have had exceptionally long, narrow wings, similar in shape to modern seabirds. The long curved teeth and sharp, upwardly curved beak, seem ideal for snaring and spearing fish over open water. Evidence suggests that *R. muensteri* would have often flown low over the water and dove or dipped into schools of fish. Part of the evidence for this behavior comes from specimens that seem to have been accidentally speared by large fish of the genus *Aspidorhynchus*, sometimes with smaller fish still in their gullet. Apparently, while the *Rhamphorhynchus* were dipping or diving from above, the sword-nosed fish were attacking the same school from below, leading to frequent collisions between predatory pterosaurs and fish. So far, only *Rhamphorhynchus* specimens have been found preserved this way, showing that it was something unique to their method of feeding that led them into such accidents (Frey & Tischlinger, 2012).

R. muensteri appears to have been a relatively long-lived species, having apparently existed relatively unchanged for over two million years. Specimens have been reported from both older (Nusplingen Formation) and younger (Moernsheim Formation) sediments relative to the Solnhofen Limestone proper. An even older specimen, found in 2002 near the village of Brunn, has since been reclassified as the species *Bellubrunnus rothgaengeri*, on the basis of unique characters such as forward-curved wing tips (Hone et al., 2012).

Specimen of R. muensteri *in the collection of the Houston Museum of Natural Science, Texas.*

Like *Pterodactylus*, *Rhamphorhynchus* seem to have undergone some significant physical changes as they grew. Flaplings had short, triangular snouts which grew longer with age, though the lower beak was just as strongly hooked in both juveniles and adults. The tail vane started out as small and oval shaped, and then grew steadily larger compared with the body. As the vane grew, it became more triangular, and in some well-preserved fossils, like the Edinburgh Specimen, the triangular vane is huge, almost as large as the head. This specimen even preserved a throat pouch similar to the one in *Pterodactylus antiquus* (Frey et al., 2003).

A Note on Nomenclature: The genus *Ornithopterus*, based on a partial specimen, most likely belongs to *Rhamphorhynchus*. *Ornithopterus* was used by Meyer for a fragmentary specimen, but later "retracted" when he realized it was a pterodactyl and identical to *Rhamphorhynchus*. This means that the well-known name *Rhamphorhynchus* would need to be changed to its older synonym. However, a search of the literature suggests that the names *Ornithopterus* Meyer 1838 was not used as valid after the year 1899, making it a candidate for nomen oblitum status and reversal of priority under ICZN Article 23.9.2. *Rhamphorhynchus* has been in prevailing use for more than 50 years as noted in many of the sources cited here, and in both the technical and popular literature. This discussion, therefore, serves to satisfy the requirements of Article 23.9.2 and designate *Ornithopterus* Meyer 1838 as a nomen oblitum. The genus name *Rhamphorhynchus* Meyer 1846 (with its type species *Pterodactylus longicaudus* Muenster 1839) should be considered valid.

Scaphognathus crassirostris

Compared with other Solnhofen pterosaurs, the species *Scaphognathus crassirostris* has a refreshingly simple taxonomic history. It was first identified by Georg August Goldfuss in 1830 from a single partial but well-preserved specimen found at Eichstätt, and, unsurprisingly, considered a new species of *Pterodactylus*. Goldfuss noted that the snout was much stouter and more robust, hence the species name, which means "fat-snouted".

Goldfuss closely reexamined the well-preserved skeleton of the first *Scaphognathus* skeleton, and made what he believed was an amazing discovery. He saw wrinkles of what he thought must be the wing membrane preserved as faint impressions in the rock, and more importantly, many short, hair-like impressions surrounding the body. Longer impressions covered the neck, forming what he thought was a horse-like mane on the animal. In some places, the little "hairs" seemed to be splitting in rays from a central point, like the simple down feathers of a bird. Though only a few quill-like structures could be seen, overall these structures covering the body looked like the feathers of an

---

"Fat Snouted Boat Jaw"

Classification:
  Class: Stem-Aves
  Order: Pterosauria
  Suborder: Rhamphorhynchoidea
  Family: Rhamphorhynchidae
  Genus: *Scaphognathus*
  Species: *Scaphognathus crassirostris*

Formations: Solnhofen

Age: Early Tithonian, 151.5 - 150.8 million years ago

Wingspan: Up to 1 meter

Main diet: Fish and other small vertebrates

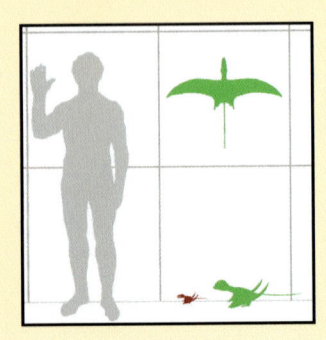

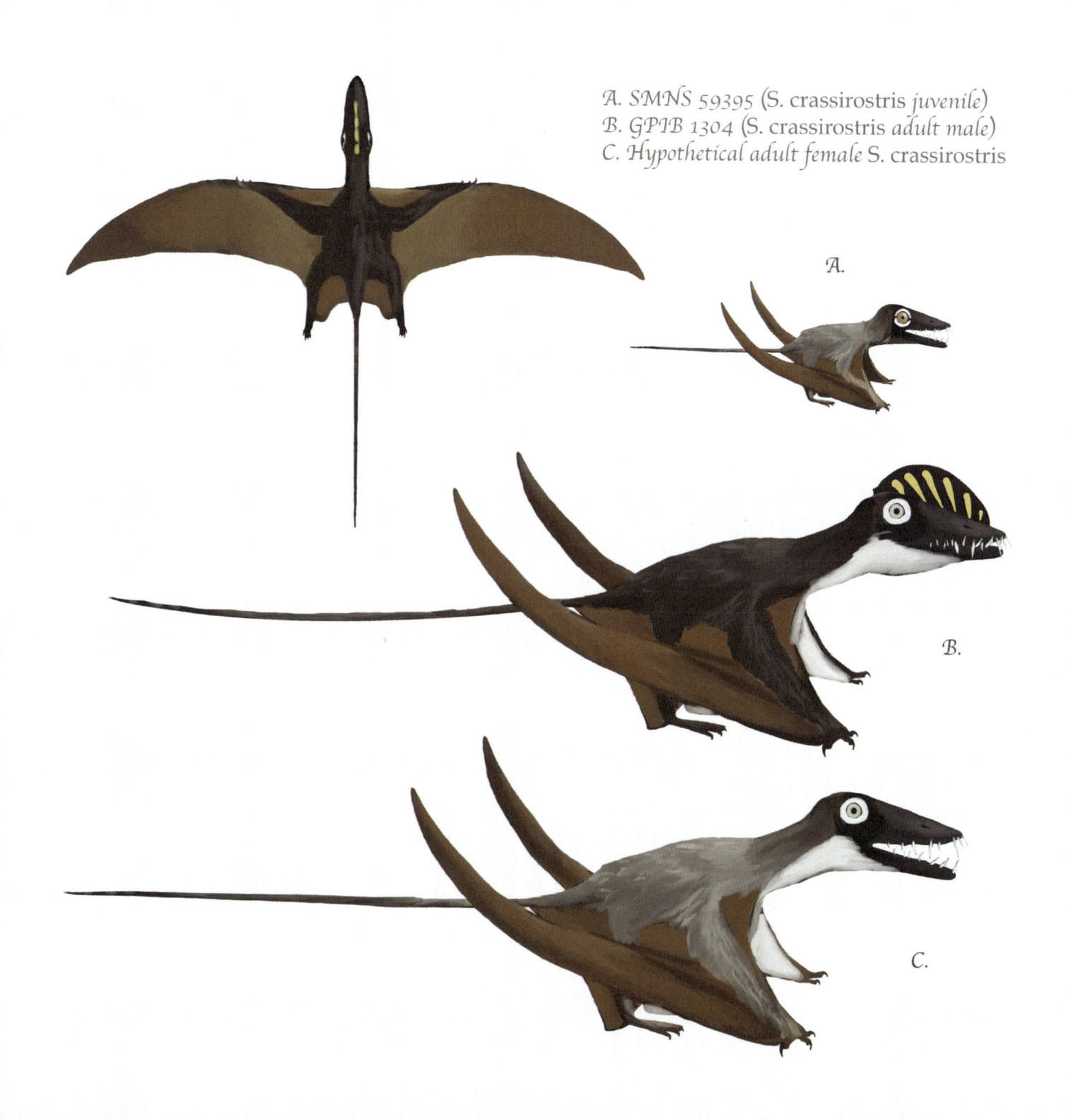

A. SMNS 59395 (*S. crassirostris* juvenile)
B. GPIB 1304 (*S. crassirostris* adult male)
C. Hypothetical adult female *S. crassirostris*

ostrich (Seeley, 1870).

Other researchers, however, were not convinced. Under a microscope, these supposed feathers became less clear, not more, and Harry Grover Seeley wrote many years later that he and other scientists of the time considered all of these traces to be simply geological formations in the rock. It was not until 2002, long after feather-like structures were well established in other pterosaurs, that Goldfuss' hypothesis was re-examined. Using ultra-violet light to reveal hard to see soft tissue, Goldfuss' interpretation was confirmed; like other pterosaurs, *Scaphognathus crassirostris* was covered in feathers, or at least feather-like filaments.

As the known diversity of pterosaurs increased, *S. crassirostris* became one of the first species to be split away from the genus *Pterodactylus*, and recognized as distinct enough for its own genus name, which means "boat-jaw" (the name was chosen by Goldfuss to contrast it with the long, slender jaws of *Pterodactylus*).

Scaphognathus crassirostris remains one of

Type specimen of S. crassirostris, *one of the first pterosaur specimens known to preserve feathers.*

the rarest species in the Solnhofen limestone, and today it is known only from three specimens: the original young adult specimen and two flaplings. As the name suggests, the skull of *S. crassirostris* was unusually short, rounded, broad and blunt compared to its closest contemporary, *Rhamphorhynchus muensteri*. The teeth were straighter, and the wings much shorter and more rounded. The possible remains of the bony core of a crest have been identified on the skull of the holotype specimen (Bennett, 2014). Some related species from China also have high, rounded crests that are present only in male specimens, and it may have been the case that *Scaphognathus* was also sexually dimorphic; however, with only one known adult specimen, this hypothesis cannot be tested yet. As for the "mane" found by Goldfuss, the long neck feathers probably formed a streamlined, aerodynamic shell instead, like the long neck feathers of many modern birds.

The rarity of *Scaphognathus* compared to other Solnhofen pterosaurs, especially juveniles, may indicate that it preferred more near-shore or inland habitats than other species. Differences in the skull, teeth, and wings, show that they lived a very different lifestyle than contemporary *Rhamphorhynchus*. The shorter wings suggest they inhabited a more landlocked environment, and short, broad wings are often associated with forest-dwelling fliers like anurognathids. In fact, a very similar species, differing mainly in details related to growth, is known from slightly older strata in China. This species, *Jianchangnathus robustus*, lived in the heavily forested ecosystem of the Tiaojishan Formation. Given the fact that no trees are known from the Solnhofen limestone, it's a possibility that these rare, short-winged fossil species like *S. crassirostris* and *Anurognathus ammoni* were occasional interlopers from the wooded inland areas of the large surrounding islands. Even if this were the case, though, *S. crassirostris* was still partially associated with water, as one specimen preserves the remains of a partially digested small fish near the jaws. If they were not primarily hunters of aquatic prey, *Scaphognathus* must have been generalists willing to find food wherever they could.

Germanodactylus rhamphastinus

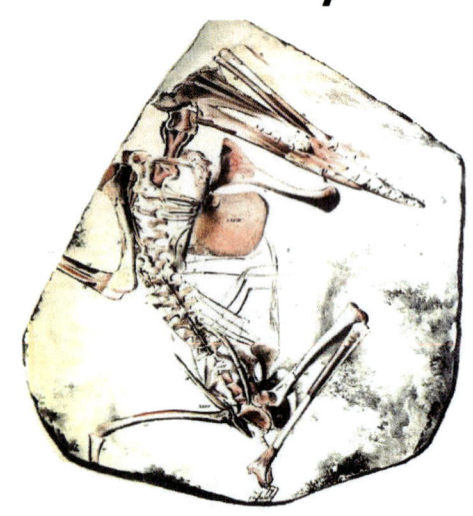

Type specimen of Pterodactylus medius, one of the first pterosaur specimens known to preserve feathers.

By the 1830s, interest in the Solnhofen pterodactyls had increased, and numerous specimens were studied and described by scientists, who often named new species for rather poor remains. One of these was *Pterodactylus medius*, an incomplete skull and skeleton. Goldfuss reported the presence of feathers or feather-like hairs on this specimen, but H.G. Seeley and others regarded these as merely geological patterns in the surrounding rock (Seeley, 1870). Goldfuss turned out to have been correct about similar features in the adult specimen of *Scaphognathus*, so it's reasonable to assume he was right in this case too. Unfortunately, the type specimen of *P. medius* was destroyed in World War II, so these features cannot be re-examined. Indeed, without the speci-

---

"Toucan-like German Pterodactyl"

Classification:
 Class: Stem-Aves
 Order: Pterosauria
 Suborder: Pterodactyloidea
 Family: Germanodactylidae
 Genus: *Germanodactylus*
 Species: *Germanodactylus rhamphastinus*

Formations: Solnhofen, Moernsheim

Age: Early - middle Tithonian, 151.5 - 150 million years ago

Wingspan: Up to 1 meter

Main diet: Fish

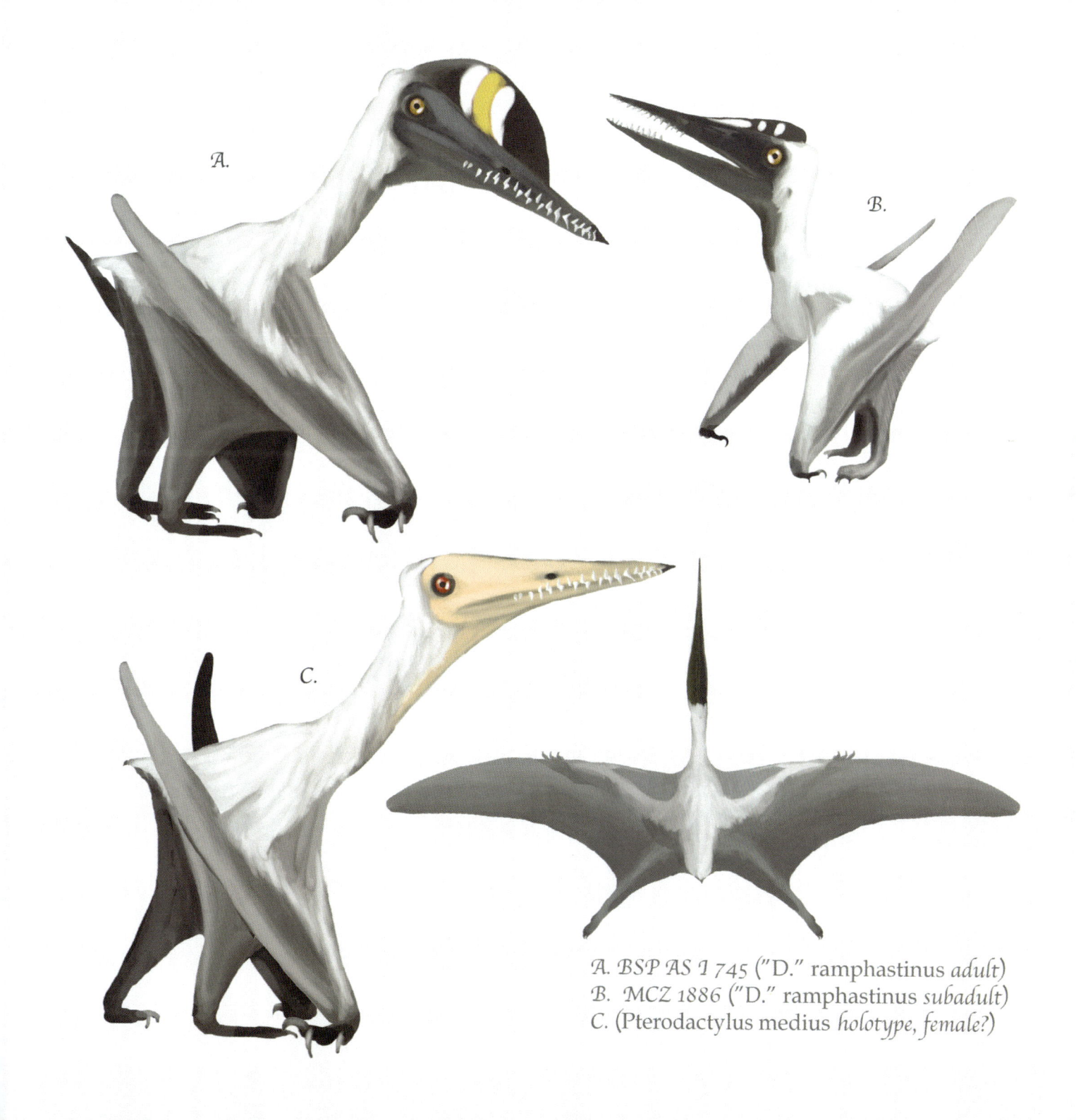

A. BSP AS I 745 ("D." ramphastinus *adult*)
B. MCZ 1886 ("D." ramphastinus *subadult*)
C. (Pterodactylus medius *holotype, female?*)

men, researchers have been reluctant to formally synonymize it with its better known, probable younger synonym, *Ornithocephalus rhamphastinus* (Bennett, 2006). Though based on a better skull, the features of *O. rhamphastinus* are a near perfect match for those reported from *P. medius*. Wellnhofer noted this in 1970, but declined to make them synonyms because the holotype of *P. medius* was gone. *Ornithocephalus rhamphastinus*, a large specimen with a prominent cranial crest, was transferred by Wellnhofer to the genus *Germanodactylus*, though it may not form a natural group with *G. cristatus*.

The head was unusually large, especially in younger, sub-adult specimens. This big head and relatively tall snout (compared to *Pterodactylus antiquus*) lent this species its name. Andreas Wagner coined the name *O. ramphastinus*, meaning "toucan-like", in reference to the large head. Later authors misinterpreted the meaning of the name as meaning "narrow beaked", and so changed the spelling to *O. rhamphastinus*. While this spelling change was not justified and was based on an incorrect assumption, it is in prevailing use, so it meets the ICZN requirement for accepting the new name (Article 32.2.3.1).

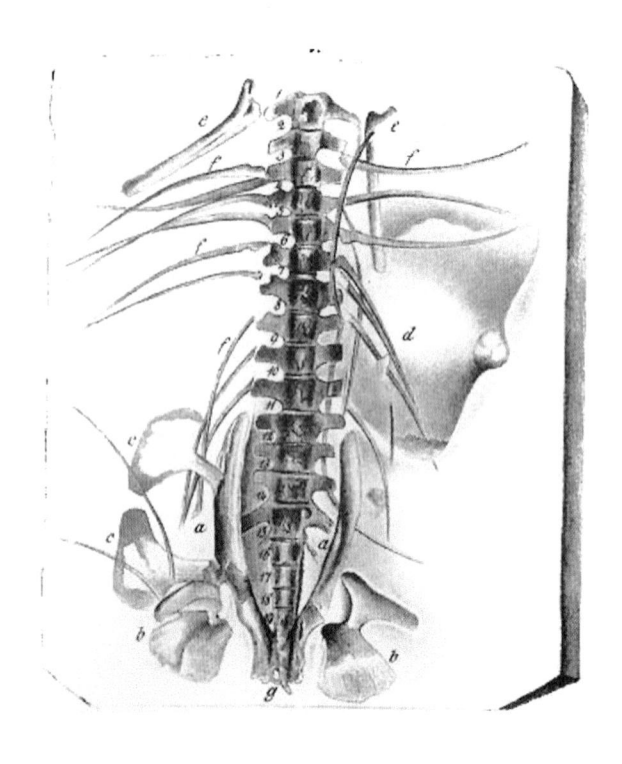

Type specimen of Pterodactylus dubius. Published in Wagner, 1851.

Overall, this species is very similar to *Pterodactylus antiquus*. The snout was long and tapering, and unlike *P. antiquus*, the jaw tips came to a point (possibly also bearing a slightly hooked beak). A prominent bony crest core is present along the middle of the skull in the adult specimen, though was apparently absent in the more mature *P. medius* specimen (possibly a result of sexual dimorphism). The teeth extended well behind the nostril and, unusually, the middle-rear teeth were larger than the front teeth. The legs were relatively long, with long, broad feet, also seen in *Pterodactylus* and *G. cristatus*. Indeed, *G. rhamphastinus* was very similar to *G. cristatus* in some respects, save for the fact that *G. cristatus* lacked teeth in the jaw tips and had a relatively smaller head as an adult.

Despite the similarities between *G. cristatus* and *G. rhamphastinus*, not all phylogenetic studies have found that they form a natural group. At least one study has suggested that *G. cristatus* is more advanced, and more closely related to the dsungaripterids. This uncertainty has led to suggestions that "*G.*" *rhamphastinus* should be placed in its own, distinct genus, which has been informally labelled "Daitingopterus", after the location in the Moernsheim formation where the holotype was found (Maisch et al., 2004).

Ardeadactylus longicollum

Like *Germanodactylus rhamphastinus*, *Ardeadactylus longicollum* bore a general similarity to *Pterodactylus*, though it differed in features of the skull and teeth. The neck was very long (hence its species name), and the head relatively small, with a long, tapered snout. The teeth were large and pointed slightly forward. There were fewer teeth, too, and they were restricted to the front third of the jaws, in front of the nostril. These differences in head and neck anatomy show that, despite their similarities, *Pterodactylus* and *Ardeadactylus* were catching different kinds of prey. It's possible that *A. longicollum* preferred catching larger fish (Bennett, 2013). Despite being found in several different formations both older and younger than Solnhofen (having a similar distribution through time to *Rhamphorhynchus muenseri*), fossils of *A. longicollum* are rare. Only five specimens have ever been found, and three of those (including the original holotype) were destroyed in World War II. An additional specimen, the holotype of *Pterodactylus longipes*, is potentially a senior synonym of *P. longicollum*, but it is too fragmentary to confidently say it belongs to this species, and not *Gnathosaurus* or *Cycnorhamphus* (Bennett, 2013).

"Long-Necked Heron Pterodactyl"

Classification:
- Class: Stem-Aves
- Order: Pterosauria
- Suborder: Pterodactyloidea
- Genus: *Ardeadactylus*
- Species: *Ardeadactylus longicollum*

Formation: Nusplingen, Solnhofen, Moernsheim

Age: Late Kimmeridgian - mid Tithonian, 152.5 - 150 million years ago

Wingspan: Up to 1.6 meters

Main diet: Large fish

No juveniles or flaplings have yet been recognized. Perhaps, like *Scaphognathus* and *Anurognathus*, this was an errant or occasional visitor to the arid Solnhofen beaches and islets, but normally leading a heron-like lifestyle among inland forest streams and ponds.

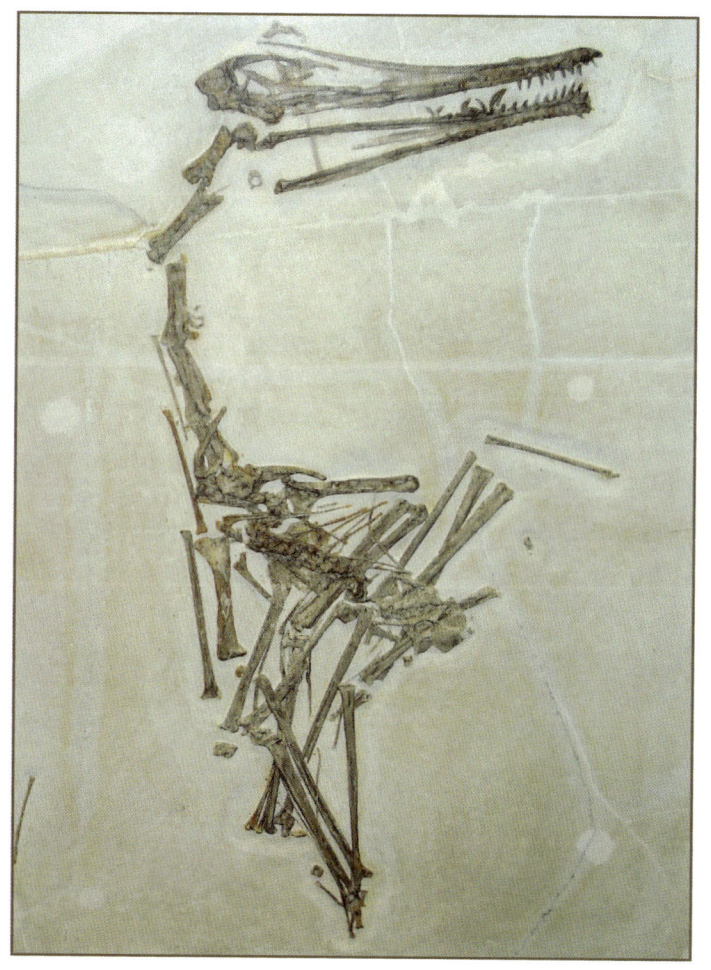

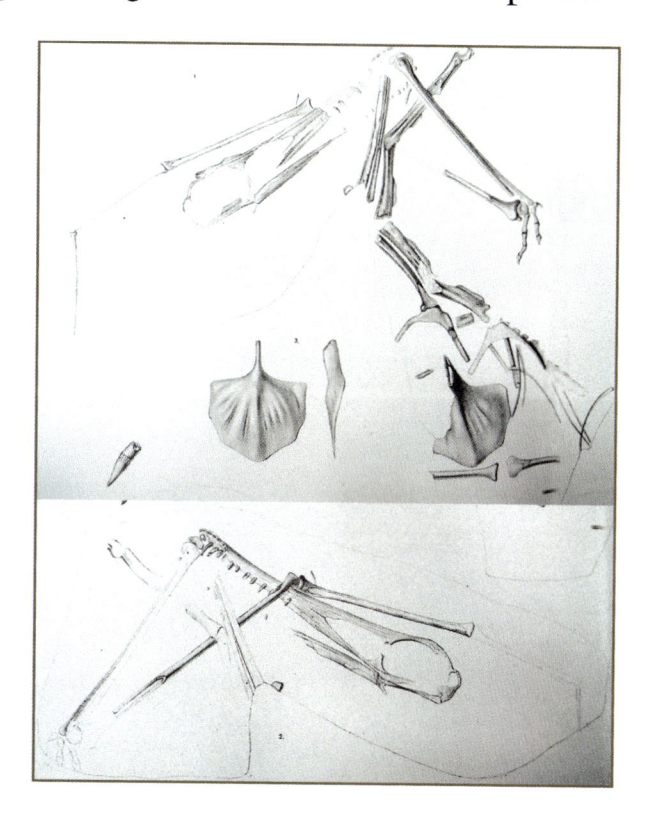

Left: Neotype specimen of A. longicollum (SMNS 56603). Currently housed in the collections of the State Museum of Natural History Stuttgart.

Right: Illustration of the A. longicollum holotype specimen slab (top) and counter-slab. From von Meyer's 'Zur Fauna der Vorwelt', volume 4. 1860.

Cycnorhamphus suevicus

Cycnorhamphus suevicus was among the rarest and most unusual of the Solnhofen pterodactyls, and of pterosaurs in general. It was first identified based on a nearly complete, but slightly disarticulated skull and skeleton originally attributed to the genus *Pterodactylus* as *P. suevicus* by Quenstedt in 1855. This original specimen came from the Kimmeridgian Nusplingen formation, not Solnhofen, but additional fossils attributed to the same species were later found there. *P. suevicus* was unusual in a number of respects, including a robust and upward-curved jaw that appeared to be mainly toothless, except for the very tip. Fifteen years later, Harry Grover Seeley re-examined the *P. suevicus*

"SWABIAN SWANBILL"

Classification:
  Class: Stem-Aves
  Order: Pterosauria
  Suborder: Pterodactyloidea
  Family: Gallodactylidae
  Genus: *Cycnorhamphus*
  Species: *Cycnorhamphus suevicus*

Formations: Nusplingen, Painten, Solnhofen

Age: Late Kimmeridgian - early Tithonian, 152.5 - 150.8 million years ago

Wingspan: Up to 2 meters

Main diet: Crustaceans & shelled mollusks?

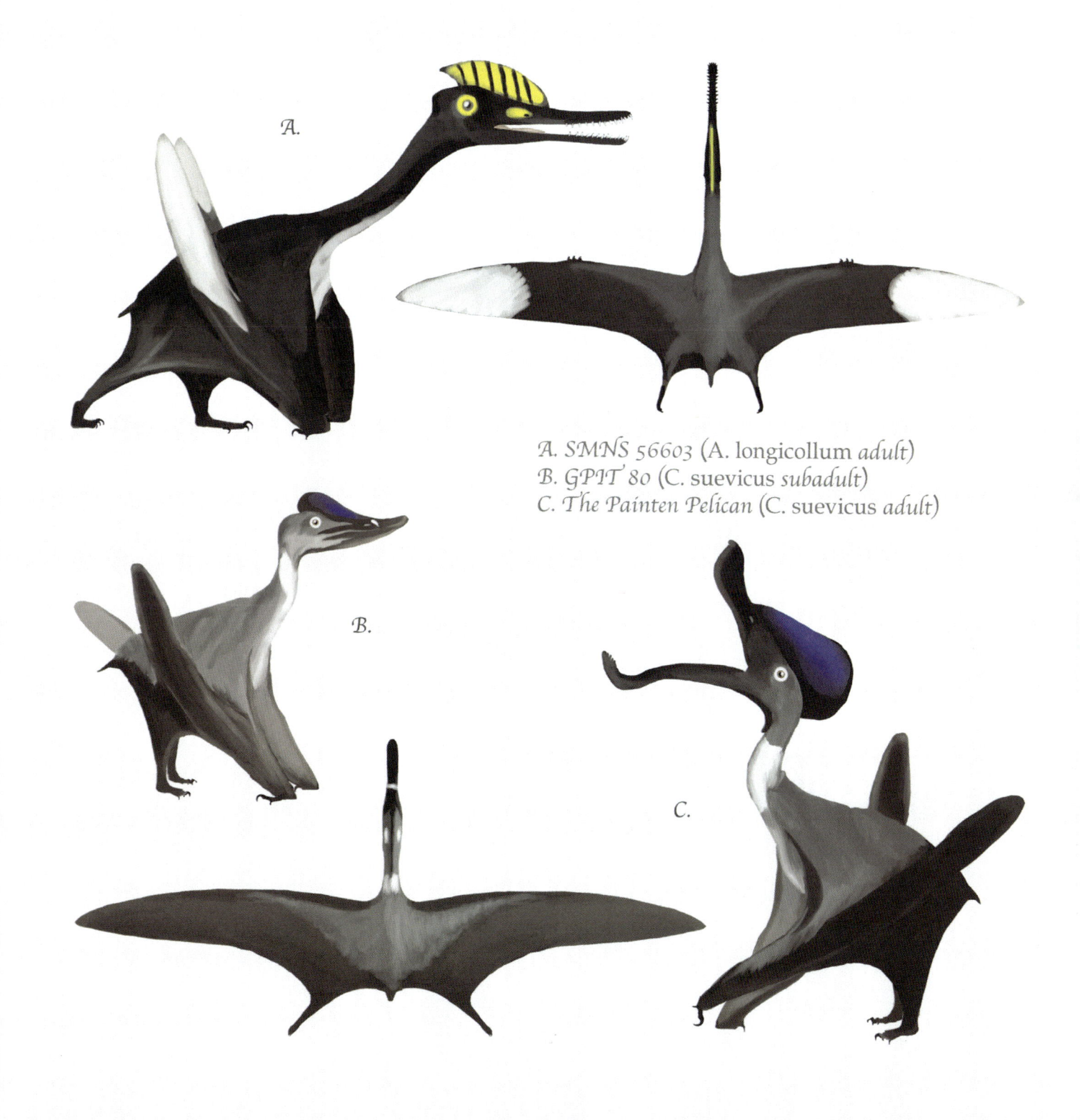

A. SMNS 56603 (A. longicollum *adult*)
B. GPIT 80 (C. suevicus *subadult*)
C. The Painten Pelican (C. suevicus *adult*)

specimens and decided that they were distinct enough to be placed in their own genus, which he named *Cycnorhamphus*, meaning "swan-beak".

C. suevicus is known from three specimens from Germany and, probably, one from France (the latter has been referred to the species *Gallodactylus canjuresensis*, but is probably synonymous with *C. suevicus* according to Bennett 1996). The holotype specimen was found in quarries of the Nusplingen formation. A second specimen was found in Eichstätt, but was held in a Munich museum collection and destroyed in a World War II bombing (this specimen was once made the holotype of a distinct species, *Pterodactylus eurychirus*).

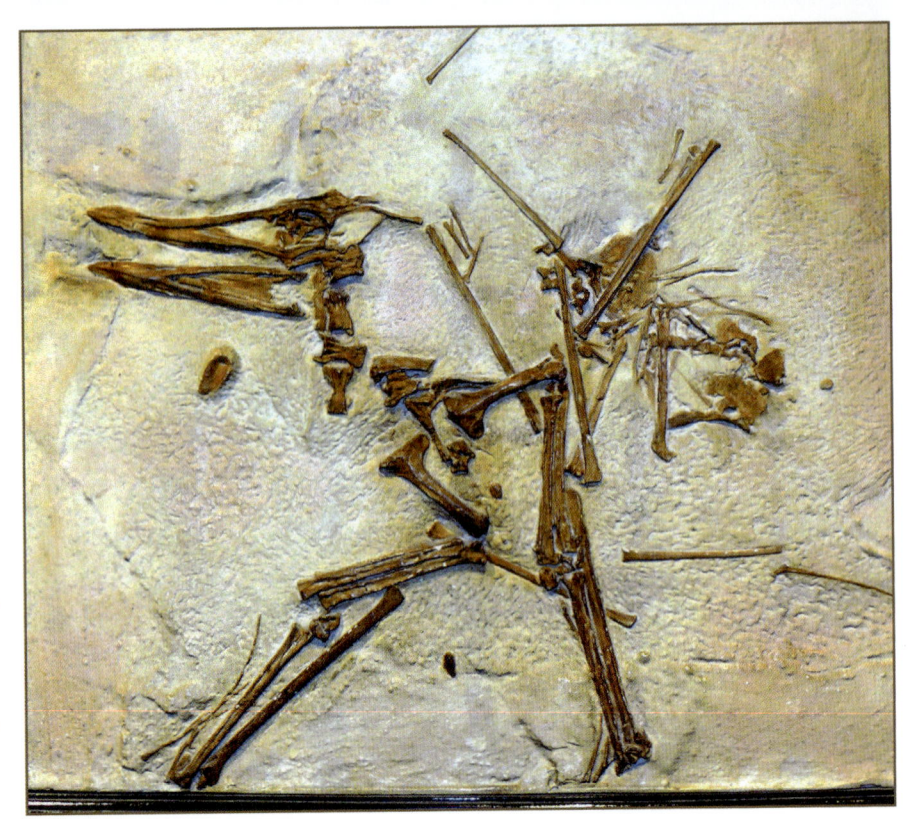

Holotype specimen of C. suevicus *(GPIT 80).*

The most remarkable specimen, from the Painten Formation (intermediate in age between the Nusplingen and the Solnhofen formations), is a skull nicknamed the "Painten Pelican". This fossil is held in a private collection, and so has not been scientifically described. It is a more mature adult specimen than the other two, and shows some startling differences, probably due to its more mature age. The teeth are even more restricted to the jaw tips than in the subadults, and the area just before the jaw tips are bent sharply away from each other, creating what looks like a prominent gap when the jaws are closed. However, impressions show that this gap was covered with some kind of soft tissue, possibly a cutting surface. The broad, pincer-like jaw tips and blunt teeth may have been used for crushing crustaceans or shelled mollusks. Impressions also show a large soft tissue crest rising from the top of the skull, and possibly connecting with an expanded "crest" at the rear portion of the skull to form one large structure. (Witton, 2013).

Germanodactylus cristatus

More than perhaps even their huge size and membranous wings, pterosaurs today are well known for their often huge and ostentatious crests. During the 19th century, however, little was known of crested pterosaurs. A few Solnhofen species preserved some indication of crests extending from the top or back of the head, but most of these were only bony cores, which were small (and now known to usually indicate much larger, soft tissue extensions) or not easily discerned.

The first pterosaur from the Solnhofen fauna widely recognized as having a prominent crest was a specimen originally thought to represent an adult *Pterodactylus kochi* (=*P. antiquus*). Discovered at the dawn of the 20th century and described by Felix Pleininger in 1901, it was later considered to be a distinct species of *Pterodactylus* by Wiman, who named it *P. cristatus* in 1925. C.C. Young made it the type specimen of the genus *Germanodactylus* in 1964. Two flapling specimens, previously described by Wellnhofer (1970) and assigned to *Pterodactylus micronyx* and *P. kochi*, were later recognized by Bennett as belonging to *G. cristatus*, due to their toothless jaw tips, a fea-

---

"CRESTED GERMAN PTERODACTYL"

Classification:
  Class: Stem-Aves
  Order: Pterosauria
  Suborder: Pterodactyloidea
  Family: Germanodactylidae
  Genus: *Germanodactylus*
  Species: *Germanodactylus cristatus*

Formations: Solnhofen

Age: Early Tithonian, 151.5 - 150.8 million years ago

Wingspan: Up to 70 centimeters

Main diet: Small fish, invertebrates

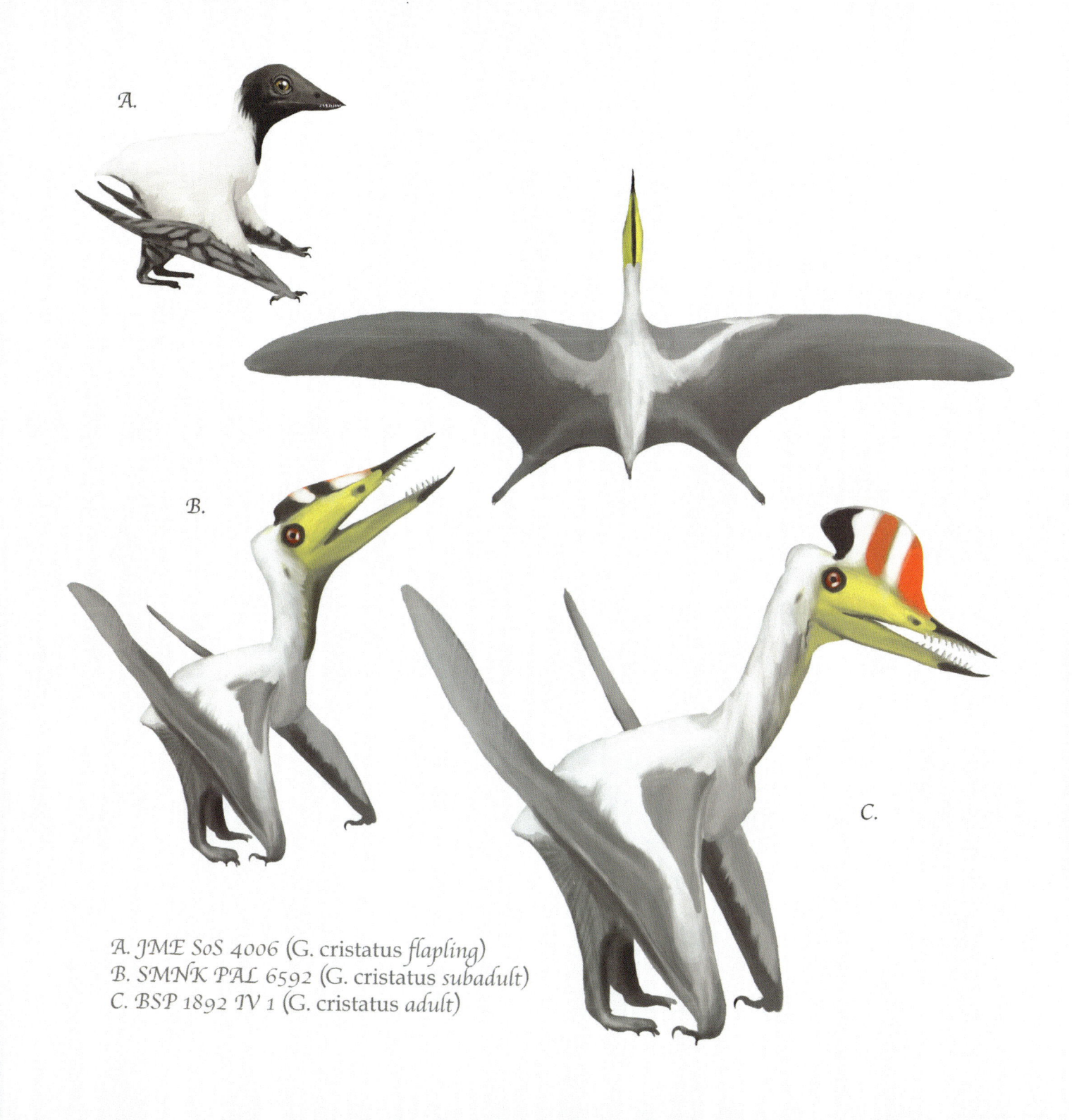

A. JME SoS 4006 (*G. cristatus* flapling)
B. SMNK PAL 6592 (*G. cristatus* subadult)
C. BSP 1892 IV 1 (*G. cristatus* adult)

ture unique to this species among Solnhofen pterodactyls.

Other than the proportionately smaller head in adults and the toothless, pointed jaw tips that characterize it, there is not much to differentiate *G. cristatus* from the contemporary *G. rhamphastinus*. The crest core of *G. cristatus* is restricted more to the middle of the skull and is slightly taller as preserved, though the two specimens are about the same size. Despite their similarities, phylogenetic studies have not always found these two species to form a group exclusive of other pterosaurs. While Andres et al. (2014) found both to be relatives close to the ctenochasmatids, some studies have found *Germanodactylus* to be an evolutionary grade closer to the dsungaripterids, a group which shares pointed, toothless jaw tips with *G. cristatus* (Maisch et al., 2004).

Specimen of a subadult G. cristatus *(SMNK PAL 6592) from the collection of the State Museum of Natural History Karlsruhe.*

Anurognathus ammoni

Until the early 1920s, the diversity of pterosaur species from Solnhofen could be fairly easily broken down into two categories - *Pterodactylus*-like forms, and *Rhamphorhynchus*-like forms. The former had generally long, narrow skulls with peg-like teeth and short tails (the bizarre ctenochasmatids had yet to be definitively recognized as pterosaurs rather than crocodilians). While the pterodactyls were beginning to seem stranger with the addition of head crests in some species, it was not until the discovery of anurognathids (and the recognition of ctenochasmatids as pterosaurians the following year) that this traditional dichotomy of basic body types was broken.

Anurognathus ammoni was difficult to interpret when first studied by Bavarian paleontologist Ludwig Heinrich Philipp Döderlein. Döderlein was compelled to try and slot it into the traditional framework of "pterodactyloid" vs. "rhamphorhynchoid" forms, and though he settled on the latter, the specimen did not quite fit. For one thing, the tail was very short, like "pterodactyloids", and the forearm very long. Then there was the skull. Though badly preserved, it was obviously extremely short and broad,

"AMMON'S FROG JAW"

Classification:
 Class: Stem-Aves
 Order: Pterosauria
 Family: Anurognathidae
 Genus: *Anurognathus*
 Species: *Anurognathus ammoni*

Formations: Solnhofen

Age: Early Tithonian, 151.5 - 150.8 million years ago

Wingspan: Up to 30 centimeters

Main diet: Flying insects

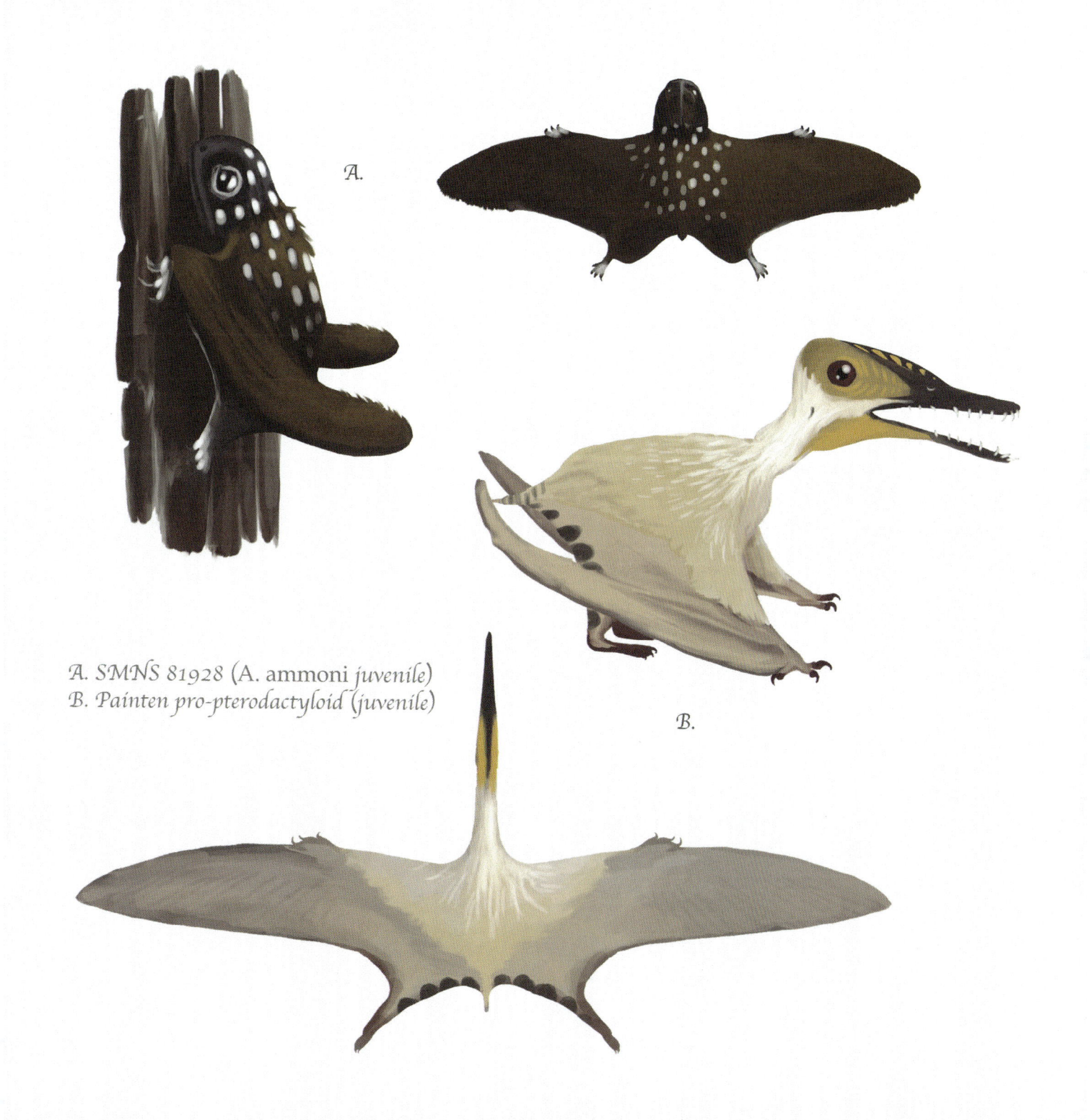

A. SMNS 81928 (*A. ammoni* juvenile)
B. Painten pro-pterodactyloid (juvenile)

with tiny, straight, needle-like teeth. This short skull, combined with the unusually short tail for its group, lent the animal its name, meaning "frog-jaw".

Subsequent discoveries of more anurognathids, like *Batrachognathus* and *Jeholopterus*, confirmed some of Döderlein's interpretations, which had been questioned by other scientists (a second specimen of *Anurognathus ammoni* itself was not found until 2007). The Chinese *Jeholopterus* in particular provided a wealth of details about anurognathid anatomy. The body was covered in a particularly dense coat of simple feathers, which extended even onto parts of the wing. The wing tips had a filament fringe on their trailing edge, which may have helped to provide these animals with quiet flight, like the frayed wing feathers of owls. The wings themselves were long but unusually broad, allowing for slow stable flight with high maneuverability. These features, combined with the short, wide face and huge, forward-facing eyes, suggest a bat-like lifestyle. *A. ammoni* probably made its living by hunting insects out over the water at night and roosting in trees during the day. Of course, the Solnhofen limestone does not preserve the kind of forest habitat anurognathids are typically found in, and the rarity of *A. ammoni* fossils suggests it was probably an accidental or occasional visitor to this seashore ecosystem.

The discovery of *Anurognathus* in 1923 marks the end, so far, of new Solnhofen pterosaur diversity. It would seem that scientists have uncovered all of the species that were habitually present there, and that any new discoveries will help flesh out the details of their growth. Indeed, most new Solnhofen species named since *Anurognathus*, like *Rhamphorhyncnhus carnegei* and *Aurorazhdarcho primordius*, have turned out to be growth stages of known species (*R. muensteri* and *Gnathosaurus subulatus*, respectively). It's a testament to the fantastic preservation possible in the Solnhofen limestone that between 1809 and 1923, a little over one hundred years, we had recognized a nearly complete fauna of pterosaurs, a group notorious for their under-representation in other geological formations.

Or, maybe, the Solnhofen has a few surprises left. In the last few years, two new species have been found from the Southern

Franconian Alb; not from the Solnhofen limestones, but from older geological formations. One, *Bellubrunnus rothgaengeri*, seems to be an evolutionary cousin or precursor of the younger *Rhamphorhynchus*. A second seems to be an evolutionary precursor to, possibly, much of the Solnhofen pterodactyloid diversity. This species was not given a name, due to the fact that the only known specimen is held in a private collection, but it shows a "transitional" mix between the characteristics typical of the Solnhofen pterodactyloid pterosaurs, and the more primitive "rhamphorhynchoid" species. Found in the Painten Formation, it has been nicknamed the "Painten Pro-pterodactyloid" (Tischlinger & Frey, 2014). Another potentially transitional species, this one from the younger Moernsheim formation, has been nicknamed "Rhamphodactylus" (Rauhut, 2013). The Painten and Moernsheim pterosaur faunas do not seem to differ too significantly from the Solnhofen fauna, and several species are shared among them (including *Pteordactylus antiquus*, *Rhamphorhycnhus muensteri*, and *Ardeadactylus longicollum*). Maybe specimens of the Painten Pro-pterodactyloid, "Rhamphodactylus", and other new but rare species are waiting yet to be found in the Solnhofen limestone.

Juravenator starki.

THEROPODS

While the Solnhofen pterodactyls possessed undeniably bird-like features, these had always been viewed mainly as coincidental or, after the advent of evolutionary biology, convergent - having evolved independently similar features due to the constraints of flight. This is still largely the case today; while most scientists consider pterosaurs to be stem-birds, they are probably a very early branch far removed from the origin of true birds.

Compsognathus longipes

It was not until 1859 that a contender for a more immediate bird ancestor was found. This specimen first turned up in the collection of a physician named Dr. Joseph Oberndorfer, who was also an amateur fossil collector. The specimen was found in an unknown quarry in the eastern region of the Southern Franconian Alb around Kelheim, and so probably part of the Solnhofen fauna, though possibly from the Painten formation (Ostrom, 1978; Reisdorf & Wuttke, 2012). Bavarian paleontologist Andreas Wagner briefly discussed and described the specimen in 1859, giving it the name *Compsognathus longipes*. He did not classify the specimen other than noting that it was a reptile ("saurier"). Wagner described *C. longipes* in more detail in a 1861 paper. In 1866, this specimen, along with the rest of Dr. Oberndorfer's collection, was purchased by the Bavarian State Collection for Paleontology and Historical Geology (Bayerische Staatssammlung für Paläontologie und historische Geologie, or BSP) in Munich.

Compsognathus longipes was discovered in Bavaria at the same time that *Hadrosaurus foulkii* was being studied by Joseph Leidy in the United States. Together,

---

"LONG-FOOTED ELEGENT JAW"

Classification:
 Class: Stem-Aves
 Order: Saurischia
 Suborder: Theropoda
 Family: Compsognathidae
 Genus: *Compsognathus*
 Species: *Compsognathus longipes*

Formation: Solnhofen? Painten?

Age: Early Tithonian, 151.5 - 150.8 million years ago

Length: Up to 1.3 meters

Main diet: Small reptiles

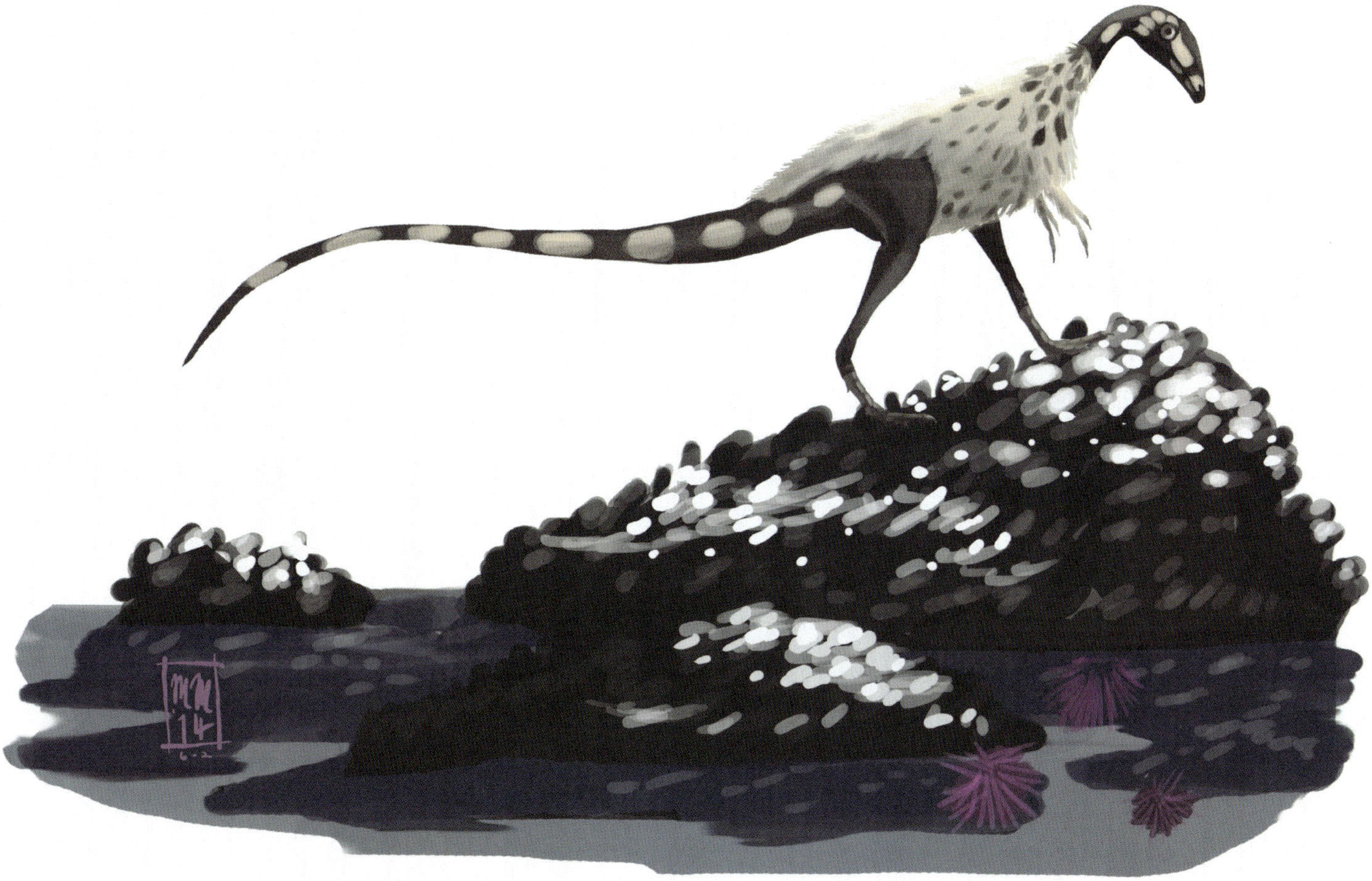

these two specimens caused a revolution in the way prehistoric life was depicted. For the first time, skeletons of dinosaurs were found that were complete enough to show that the forelimbs were much shorter than the hind limbs, showing that some birdlike reptiles, both tiny and gigantic, walked bipedally. Huxley cited this mounting evidence to suggest that the oldest supposed bird fossils, a series of footprints from Massachusetts, may actually have been made by "birdlike reptiles" (basically, stem-birds) such as dinosaurs or *Compsognathus* (which was not recognized as a type of dinosaur at the time).

To date, *Compsognathus* is represented by only two or three fossils (the third, a fragmentary hind limb, may well belong to a different kind of small theropod dinosaur). The second, a specimen from France (now held in the collections of the Museum d'Eistoire Naturelle in Nice), was found in 1971 and originally identified as a new species (*C. corallestris*), but is probably just a more mature *C. longipes*. The Munich specimen of *C. longipes* is probably immature if not a juvenile.

The Munich *C. longipes* specimen preserves the bones of a small stem-lizard (*Bavarisaurus macrodactylus*) in its stomach, and so these small theropods were probably mainly carnivorous. *Bavarisaurus* had proportions similar to modern fast-running, ground-dwelling lizards, and so *C. longipes* must have been a fast and agile hunter of small prey. The forelimbs of *C. longipes* don't seem to be adapted to capturing prey, even though they were fairly robust. The short, fairly straight claws may have been used primarily for digging (Ostrom, 1978).

Because *C. longipes* has been known from almost complete specimens since the mid-19th century, ideas about its basic life appearance have not changed much over the intervening years. One area of its anatomy that has caused considerable confusion and controversy, though, is the hand. In both known specimens, the hand is disarticulated, and scientists have disagreed about its basic structure. The Munich specimen appeared to preserve only four claws and a total of two fingers, plus a tiny vestige of a clawless third finger. The Nice specimen supposedly preserved three complete digits, possibly even with a robust third finger. What's more, the specimen preserves a long, fin-like impression of soft tissue extending out from the end of the forelimb, leading to an odd, short-lived hypothesis that *Compsognathus* had flippers (Bidar et al., 1972; Halstead, 1975). However, re-study of both specimens has shown that only two functional fingers were likely present on each of their hands, and that the supposed evidence for flippers was actually fossilized ripples (Ostrom, 1978; Peyer, 2006).

Cast of the C. longipes *holotype.*

Archaeopteryx lithographica

The fossil record of "birdlike reptiles" was well established by the mid-1800s, in the form of pterosaurs and various types of dinosaurs, including *Compsognathus*. But the record of what would be considered actual birds at the time was paltry. Fossilized bird skeletons were expected to be present in Jurassic rocks, even in the early 19th century, thanks to sets of New England footprints reported by American paleontologist Edward Hitchcock in 1836 (Switek, 2010). Jokingly referred to as "Noah's raven" tracks or turkey tracks by local farmers and masons (McCarthey, 2010), these petrified footprints dated to the Triassic period. Hitchcock noted that while some of the tracks definitely came from birds, many appeared to be

Cast of the London Specimen of Archaeopteryx lithographica.

"LITHOGRAPHIC ANCIENT WING"

Classification:
  Class: Stem-Aves
  Order: Saurischia
  Suborder: Theropoda
  Family: Archaeopterygidae
  Genus: *Archaeopteryx*
  Species: *Archaeopteryx lithographica*

Formations: Solnhofen

Age: Early Tithonian, 151.5 - 150.8 million years ago

Length: Up to 50 centimeters

Main diet: Small reptiles, invertebrates

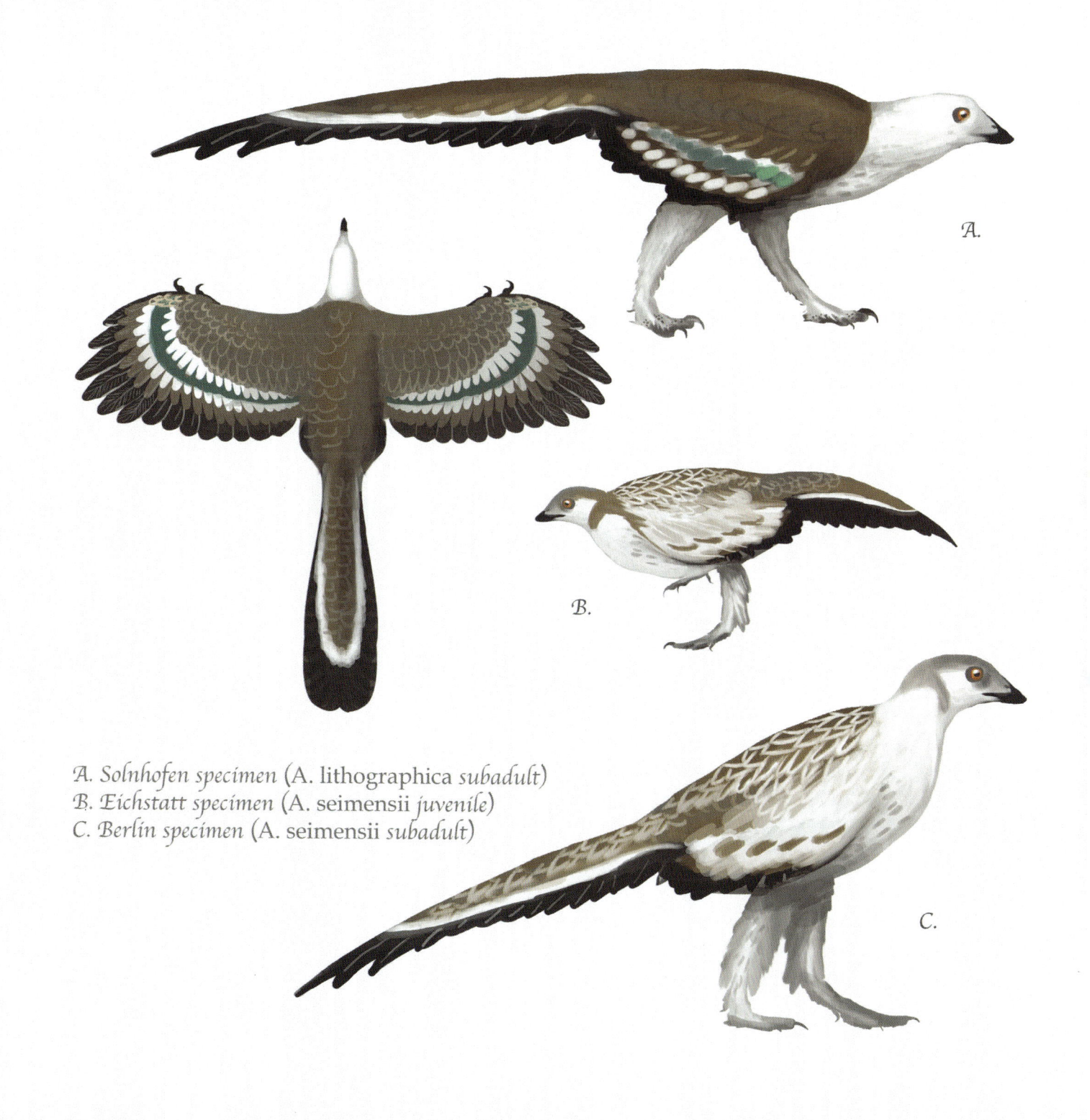

A. Solnhofen specimen (*A. lithographica* subadult)
B. Eichstatt specimen (*A. seimensii* juvenile)
C. Berlin specimen (*A. seimensii* subadult)

quadrupedal, and left traces of long tails, and so must have been made by an animal sharing traits with both birds and marsupials (like some at the time thought of the pterodactyls). Hitchcock called these animals "ornithoid marsupialoids" (Hitchcock, 1858). These tracks are now known to belong to dinosaurs (not coincidentally, the earliest reconstructions of bipedal dinosaur skeletons several decades later imagined them as very kangaroo-like in posture and gait).

Despite a presumed footprint record of nearly modern-looking birds dating back to the Triassic, the oldest known bird skeletons known leading up to the discovery of *Archaeopteryx* were those of recently extinct flightless birds such as the Moa and the Elephant Bird, which, despite their larger size, were not very different from modern flightless birds. So it came as quite a surprise when the neotype skeleton of *Archaeopteryx lithographica* was found in Jurassic rocks, displaying such a bizarre mixture of primitive and modern traits.

The name *Archaeopteryx lithographica* was originally applied to an isolated fossil wing feather found in 1861, but was later transferred to the actual skeleton discovered later that year. However, neither the neotype skeleton nor the feather was the first known fossil protobird specimen. A fragmentary skeleton, complete with faint and hard-to-interpret feather impressions, was discovered in 1855 near Kelheim. Meyer

Solnhofen Specimen of Archaeopteryx lithographica.

identified this fossil as a new species of *Pterodactylus, P. crassipes*. This mistake was not corrected until American paleontologist John Ostrom recognized the specimen as an *Archaeopteryx* in 1970, and the older species name *A. crassipes* was officially rejected in favor of the newer but universally-used name *A. lithographica*.

The first skeleton of *Archaeopteryx* to be recognized as such was unearthed in a quarry near Langenaltheim, in western Bavaria. It was given to a physician, Dr. Karl Häberlein, possibly as payment for services. Häberlein sold it to the Natural History Museum in London for 700 pounds, where it is housed today, known as the London specimen (Chiappe, 2007). The London specimen was described by British paleontologist Sir Richard Owen in 1863 and was named *Archaeopteryx macrura*. Charles Darwin was quick to add a mention of the London specimen to his fourth edition of *On the Origin of Species* in 1866, not necessarily as evidence for evolution, but for the fact that the fossil record was still capable of producing surprising discoveries and filling in gaps in lineages. "Not long ago", he wrote, "palaeontologists maintained that the whole class of birds came suddenly into existence during the eocene period; but now we know, on the authority of Professor Owen, that a bird certainly lived during the deposition of the upper greensand; and still more recently, that strange bird, the *Archeopteryx* [sic], with a long lizard-like tail, bearing a pair of feathers on each joint, and with its wings furnished with two free claws, has been discovered in the oolitic slates of Solnhofen. Hardly any recent discovery shows more forcibly than this how little we as yet know of the former inhabitants of the world."

Archaeopteryx siemensii

The London specimen of *Archaeopteryx* (colloquially called "urvogel"), along with the original feather, remained the only Solnhofen bird fossils known to science until the late 1870s. A more complete specimen, the first to include the skull, was found in a quarry near Eichstätt in 1875, and changed hands many times before eventually coming to the attention of the scientific community. The fossil was found in the Blumenberg quarry by a local farmer, Jakob Niemeyer. Niemeyer sold the specimen to a local innkeeper named Johann Dörr the following year. Dörr, in turn, sold it to Ernst Otto Häberlein, the son of Dr. Karl Häberlein, who had sold the first specimen to the London museum several years earlier (Chiappe, 2007). Häberlein knew that this fossil, though not yet fully prepared out of its limestone slab, was more complete than the one his father had sold, and its completeness and rarity meant it would fetch a high price. He began to show it to various potential buyers, and most notably displayed it at meeting of the Bavarian Academy of Sciences on May 5th, 1877. However, he had apparently priced the specimen too high at 26,000 marks, and museums repeatedly turned down his offers of sale. Wealthy

---

"SIEMENS' ANCIENT WING"

Classification:
  Class: Stem-Aves
  Order: Saurischia
  Suborder: Theropoda
  Family: Archaeopterygidae
  Genus: *Archaeopteryx*
  Species: *Archaeopteryx siemensii*

Formations: Solnhofen

Age: Early Tithonian, 151.5 - 150.8 million years ago

Length: Up to 35 centimeters

Main diet: Small reptiles, invertebrates

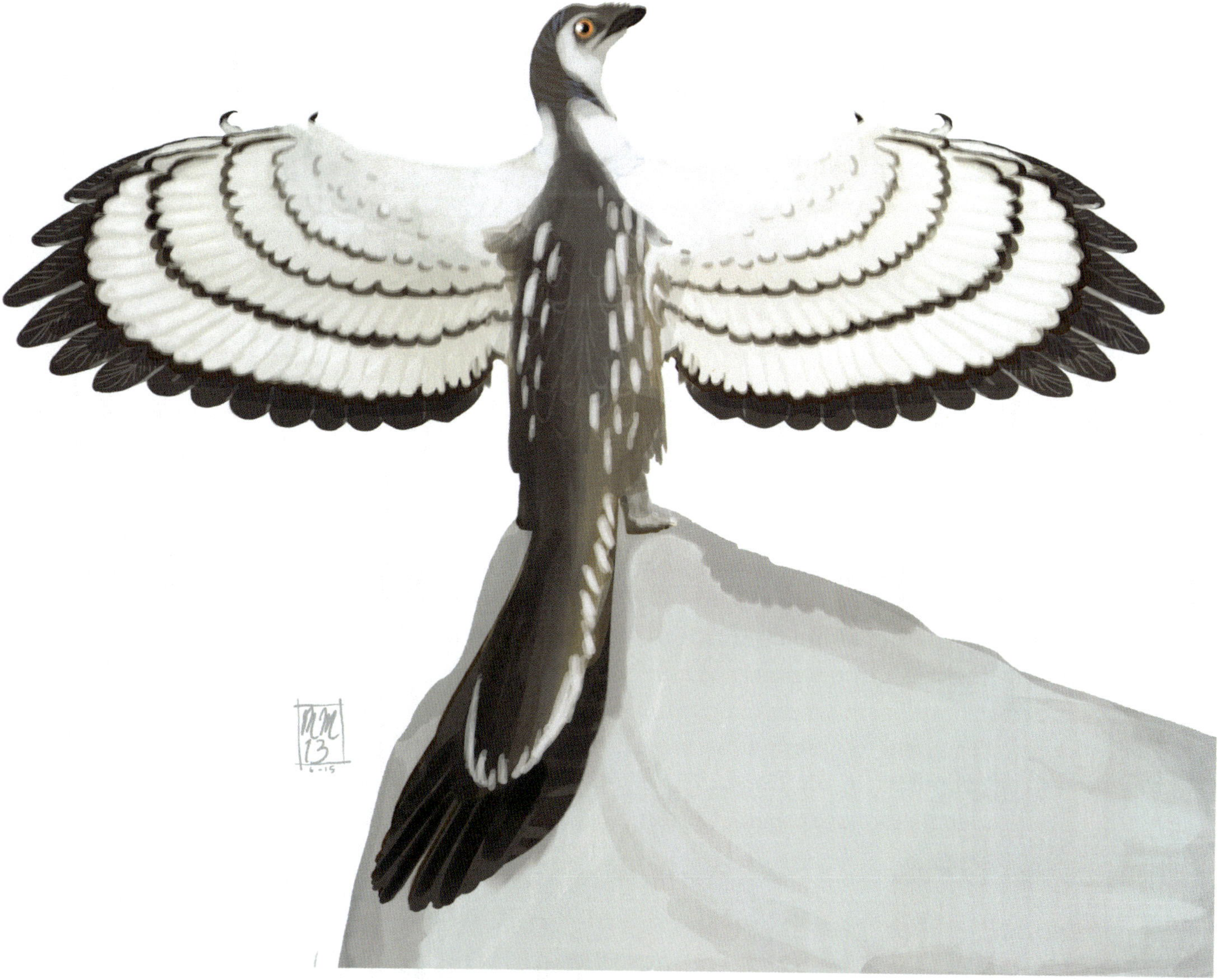

German industrialist Ernst Werner Siemens, founder of the Siemens corporation, personally purchased the specimen from Häberlein in 1880 for 20,000 marks, and shortly thereafter, sold the specimen to the Humboldt Museum in Berlin (Switek, 2014). From that point on, the fossil was known as the Berlin Specimen. After being fully prepared and studied, a scientific description was published four years later by Wilhelm Dames, who continued to study the specimen. In 1897, Dames concluded that the Berlin specimen represented a species distinct from the London specimen, and named it *Archaeopteryx siemensii*.

The species status of the Berlin and other urvogel specimens has remained controversial, and has yet to be thoroughly studied. Several more specimens were found during the 20th and 21st centuries, many of which were assigned to new species and/or genera. Much of this variation is probably due to age differences. However, the researchers who studied the tenth specimen, known as the Thermopolis specimen, concluded that at least two distinct species can be recognized, based on features such as the presence (in the London and Solnhofen specimens) or absence (in the Berlin, Thermopolis, and other specimens) of strong flexor tubercles on the claws, and the size and shape of the ischium (Meyr et al., 2007). This most recent result is followed here, pending further study of urvogel diversity.

In 1917, Petronievics suggested that *A. siemensii* was distinct enough to warrant not only its own species but its own genus, which he named *Archaeornis*. This has not been followed by subsequent researchers, but the name should be considered available if future studies suggest *A. siemensii* and *A. lithographica* are not each other's closest relatives.

With eleven specimens scientifically described, many of them reasonably complete, it is surprising that so many restorations and reconstructions of urvogel, especially in the popular press but also occasionally in the scientific literature, have gotten the details of their appearance so wrong. The most common mistake may be depicting three free wing claws. Like birds, the wing of *Archaeopteryx* was anchored to, and bound up with, the digits of the hand. Some fossils preserve the third finger crossing over the second, implying that it was probably free of the body of the wing, but the wings would not have little "hands" poking out of them as often depicted. Similarly, the head and neck are often depicted as bare or even scaly, likely to emphasize the hybrid bird-reptile nature of these animals, but there is no fossil evidence suggesting a naked head. Scales on the head, in particular, are extremely unlikely, since more primitive species, like *Anchiornis huxleyi*, show that early avialans had fully-feathered heads and that this was probably the ancestral condition.

The most recently described, eleventh specimen of *Archaeopteryx* (part of the private collection of Dr. Burkhard Pohl) has the finest feather preservation of any specimen. Based on this specimen, as well as the Berlin specimen, it is possible to describe the outward appearance of *Archaeopteryx* in great detail (Foth et al., 2014). Long, pennaceous feathers covered the neck and probably at least parts of the head. The wing feathers were long but relatively narrow, and included an overlying layer of covert feathers about half as long as the flight feathers. The body was also covered in pennaceous feathers. Long tail feathers extended the full length of the tail, two per vertebrae. Very long feathers were present on the tibia, forming "feather trousers" similar to those of hawks. Shorter feathers covered the tarsus and may even have extended onto the toes as in related protobirds like *Anchiornis*.

Eichstätt specimen of Archaeopteryx siemensii.

Could *Archaeopteryx* fly? Probably not well, though it had most of the necessary equipment. Contrary to earlier studies, the rachis (central quill) of the wing feathers was not too thin to bear weight, but was in line with the thickness found in modern birds (Foth et al., 2014). The feathered surfaces of *Archaeopteryx*, including the tail, were at least theoretically capable of generating and sustaining lift, with the tail providing a means for a low take-off velocity (Meseguer et al., 2012). Though some studies have found that the wing of *Archaeopteryx* and other stem-birds with similar anatomy (like *Deinonychus*) could not be raised above the horizontal due to the position of the shoulder joint, this may have been based on inaccurate comparisons; in many species, the shoulder socket was apparently higher up on the torso than usually thought, allowing at least some degree of flapping (Parsons & Parsons, 2009; Agnolin & Novas, 2013). However, *Archaeopteryx* lacked a key attachment point for shoulder muscles between the shoulder and the upper arm bone. Unless other muscles were enlarged and modified to compensate for this, *Archaeopteryx* would not have been capable of a modern-style flight stroke (Parsons & Parsons, 2009). It's likely that *Archaeopteryx* could execute short bursts of weak flapping flight, to extend leaps or evade predators, but could not truly fly.

Right: Berlin specimen of Archaeopteryx, *holotype of* A. siemensii.

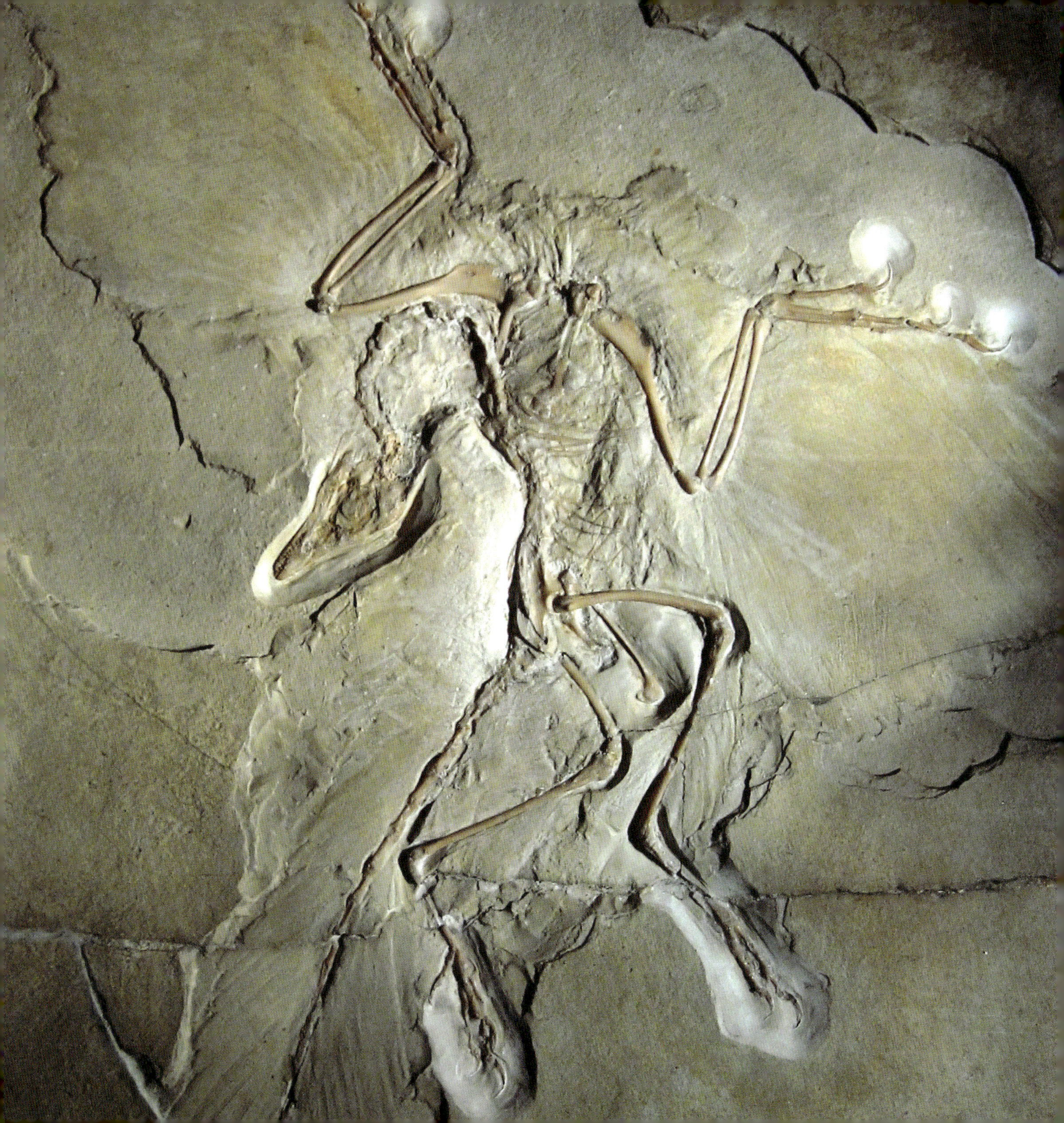

FUTURE FINDS

Our understanding of the diversity of stem-birds from Solnhofen, the cradle of their discovery, continues to improve. Further research into the field of ontogeny, or how these creatures changed as they grew, will no doubt continue to revise our thinking about which pterosaur flaplings should be matched with which adults, and exactly how many species of *Archaeopteryx* coexisted in this ancient archipelago. And, no doubt, entirely new species and new specimens of known species will eventually be found, some helping to answer longstanding questions. How did the bizarre *Cycnorhamphus* use its jaws? What was the body of an adult *Gnathosaurus* like (and how similar was it to the skeleton of *Aurorazhdarcho*?). Did the crests of the pterodactyls here vary by sex as they do in Chinese species? Why did *Rhamphorhynchus* so often get into fatal collisions with spear-nosed fish? Exactly how well could *Archaeopteryx* fly, if at all? Answers, as well as new mysteries, lie patiently waiting to be uncovered in the layers of the *Solnhofen* limestone.

GLOSSARY

Caruncle
Any fleshy outgrowth that grows normally from the skin, usually functioning in display. Examples include wattles, combs, snoods, and lappets.

Flapling
A baby pterosaur, named in reference to the hypothesis that pterosaurs were capable of powered, flapping flight soon after hatching. The term "flapling" was coined by David Unwin in 2003 and has sometimes been adopted by other researchers, mainly in popular literature and blogs (Unwin, 2006; Hing, 2011).

Holotype
The specimen to which the name of a species is attached. Used as the basis for comparison allowing other specimens to be classified, or not, in the same species.

Lappet
A soft tissue display structure formed by folded skin caruncles.

Neotype
The specimen to which the name of a species is attached, but which has been transferred from the original holotype to a new specimen, e.g. because the original holotype was lost, or it was deemed to be inadequate.

Patagium
Membrane of muscle and skin forming some or all of a wing in birds and pterosaurs.

Pennaceous
A type of feather with a central quill (rachis) and a vane made up of smaller barbs and barbules; the typical kind of feather found in modern birds, as opposed to plumulaceous feathers like down.

Pterodactyl
Originally a generic term for all "flying reptile" fossils, many modern scientists prefer to restrict its use to the genus *Pterodactylus*.

Pycnofibres
Bristly or downy body covering of pterosaurs, possibly a type of primitive feather (Kellner et al., 2010).

Stem-bird
Any animal more closely related to birds than to crocodilians, but which is not itself a true bird (member of the crown clade Aves).

Urvogel
Vernacular German name for *Archaeopteryx*, meaning "first bird".

Taxonomic Index

- *Panaves* Gauthier 2001
  - *Pterosauria* Kaup 1831
    - *Anurognathidae* Nopcsa 1928
      - *Anurognathus* Doederlein 1923
        - - *A. ammoni* Doederlein 1923
    - *Rhamphorhynchidae* Seeley 1870
      - *Rhamphorhynchinae* Seeley 1870
        - *Rhamphorhynchus* Meyer 1846
          Synonyms:
          Ornithopterus Meyer 1838 (nomen oblitum)
          Odontorhyncnus Stolley 1936
          Pteromonodactylus Teryaev 1967
          - - *R. muensteri* (Goldfuss 1831)
            Synonyms:
            Ornithocephalus giganteus? Oken 1819
            Pterodactylus grandis? Cuvier 1824
            P. longicaudus Muenster 1839
            P. lavateri Meyer 1838
            P. secundarius Meyer 1843
            P. gemmingi Meyer 1846
            R. suevicus Fraas 1855
            P. hirundinaceus Wagner 1857
            R. curtimanus Wagner 1858
            R. longimanus Wagner 1858
            R. meyeri Owen 1870
            R. phyllurus Marsh 1882
            R. longiceps Woodward 1902
            R. kokeni Plieninger 1907
            R. megadactylus Koenigswald 1931
            Odontorhynchus aculeatus Stolley 1936
            R. carnegiei Koh 1937
      - *Scaphognathinae* Seeley 1870
        - *Scaphognathus* Wagner 1861
          Synonyms:
          Pachyrhamphus Fitzinger 1843 (preoccupied)
          Brachytrachelus Giebel 1850 (preoccupied)
          - - *S. crassirostris* (Goldfuss 1830)
    - *Pterodactyloidea* Plieninger 1901
      - *Archaeopterodactyloidea* Kellner 1996
        - *Pterodactylus* Cuvier 1809
          Synonyms:
          Ornithocephalus Soemmering 1812
          Pterotherium Fischer 1813
          Macrotrachelus Giebel 1852
          Diopecephalus Seeley 1871
          - - *P. antiquus* (Soemmering 1812)
            Synonyms:
            P. longirostris Cuvier 1819
            P. crocodilocephaloides Ritgen 1826
            O. kochi Wagner 1837
            P. mayeri Muenster 1842
            P. scolopaciceps Meyer 1850
            P. spectabilis Meyer 1861
            P. westmani Wiman 1925
            P. cormeranus Doederlein 1929
        - *Germanodactylus* Young 1964
          - - *G. rhamphastinus* (Wagner 1831)
            Synonyms:
            Pterodactylus medius? Muenster 1831
            P. dubius? Meyer 1843
            P. propinquis Wagner 1858
          - - *G. cristatus* (Wiman 1925)
        - *Ardeadactylus* Bennett 2013
          - - *Ardeadactylus longicollum* (Meyer 1854)
            Synonyms:
            Pterodactylus longipes? Muenster 1836

P. vulturinus Wagner 1858
P. dubius? Meyer 1843
Cycnorhamphus fraasii Seeley 1891
- *Cycnorhamphus* Seeley 1870
  Synonym:
  Gallodactylus Fabre 1976
- - *C. suevicus* (Quenstedt 1855)
  Synonyms:
  Pterodactylus wuerttembergicus Quenstedt 1854 (nomen oblitum)
  P. eurychirus Wagner 1858
G. canjuresensis Fabre 1976
- *Ctenochasmatidae* Nopsca 1928
  - *Gnathosaurinae* Unwin 1992
    - *Gnathosaurus* Meyer 1834
    Synonym:
    Aurorazhdarcho? Frey et al. 2011
    - - *G. subulatus* Meyer 1834
    Synonyms:
    P. longipes? Muenster 1836
    P. micronyx? Meyer 1856
    P. pulchellus Meyer 1861
    G. multidens Walther 1904
    A. primordius? Frey et al. 2011
  - *Ctenochasmatinae* Nopsca 1928
    - *Ctenochasma* Meyer 1852
    Synonym:
    Ptenodracon? Lydekker 1888
    - - *C. elegans* (Wagner 1861)
    Synonyms:
    Ornithocephalus brevirostris? Soemmering 1817
    C. gracile Oppel 1862
Theropoda Marsh 1881
  Coelurosauria Huene 1914
    Compsognathidae Cope 1875
      - *Compsognathus* Wagner 1859
      - - *Compsognathus longipes* Wagner 1859

Synonym:
C. corralestris Bidar et al. 1972
Archaeopterygidae Huxley 1871
- *Archaeopteryx* Meyer 1861 (conserved name)
  Synonyms:
  Griphosaurus Wagner 1862 (rejected name)
  Griphornis Woodward 1862 (rej. name)
  Archaeornis Petronievics 1917
  Wellnhoferia Elzanowski 2001
- - *A. lithographica* Meyer 1861 (cons. name)
  Synonyms:
  Pterodactylus crassipes Meyer 1857 (rej. name)
  Griphos. problematicus Wagner 1862 (rej. name)
  Griphor. longicaudatus Woodward 1862 (rej. name)
  A. macrura Owen 1863 (rej. name)
  A. oweni Petronievics 1921 (rej. name)
  W. grandis Elzanowski 2001
- - *A. siemensii* Dames 1897
  Synonyms:
  A. recurva Howgate 1984
  A. bavarica Wellnhofer 1993

BIBLIOGRAPHY

- Agnolin, F., & Novas, F. E. (2013). *Avian Ancestors: A Review of the Phylogenetic Relationships of the Theropods* Unenlagiidae, Microraptoria, Anchiornis *and* Scansoriopterygidae. New York: Springer.

- Andres, B., Clark, J., & Xu, X. (2014). The earliest pterodactyloid and the origin of the group. *Current Biology*, 24(9): 1011-1016.

- Barrett, P. M., Butler, R. J., Edwards, N. P., & Milner, A. R. (2008). Pterosaur distribution in time and space: an atlas. *Zitteliana*, B28: 61-107.

- Barthel, K. W. (1970). On the deposition of the Solnhofen lithographic limestone (Lower Tithonian, Bavaria, Germany). *Neues Jahrbuch für Geologie und Paläontologie-Abhandlungen*, 135, 1-18.

- Barthel, K. W., Swinburne, N.H.M., and Conway Morris, S. (1990). *Solnhofen: A Study in Mesozoic Palaeontlgy*. Cambridge University Press, Great Britain.

- Bennett, S. C. (1995). A statistical study of *Rhamphorhynchus* from the Solnhofen Limestone of Germany: year-classes of a single large species. *Journal of Paleontology*, 69(3): 569-580.

- Bennett, S. C. (2007). A review of the pterosaur *Ctenochasma*: taxonomy and ontogeny. *Neues Jahrbuch für Geologie und Paläontologie-Abhandlungen*, 245(1): 23-31.

- Bennett, S. C. (2013). New information on body size and cranial display structures of *Pterodactylus antiquus*, with a revision of the genus. *Paläontologische Zeitschrift*, 87(2): 269-289.

- Bennett, S. C. (2014). A new specimen of the pterosaur *Scaphognathus crassirostris*, with comments on constraint of cervical vertebrae number in pterosaurs. *Neues Jahrbuch für Geologie und Paläontologie-Abhandlungen*, 271(3): 327-348.

- Bidar, A., Demay, L., & Thomel, G. (1972). *Compsognathus corallestris*, une nouvelle espèce de dinosaurien théropode du Portlandien de Canjuers (Sud-Est de la France). *Annales du Muséum d'Histoire Naturelle de Nice*, 1: 9–40.

- Bottjer, D. J. (Ed.). (2002). *Exceptional fossil preservation: a unique view on the evolution of marine life.* Columbia University Press.

- Broili, F. (1938). Beobrachtungen an *Pterodactylus*. *Sitzungsberichte - Bayerische Akademie der Wissenschaften, Mathematisch-Naturwissenschaftliche*: 139-154.

- Chiappe, Luis M. (2007). *Glorified Dinosaurs*. Sydney: University of New South Wales Press.

- Cuvier, G. (1809). Sur le squelette fossile d'un *Reptile Volant* des environs d'Aischstedt, que quelques naturalistes ont pris pour un oiseau, et dont nous formons un genre de *Sauriens*, sous le nom de *Petro-Dactyle*. *Annales du Muséum d'Histoire Naturelle*, v. 13.

- Cuvier, G. (1812). *Recherches sur les Ossemens Fossiles* (1st ed.).

- Delair, J. B., and Sarjeant, W.A.S. (2002). The earliest discoveries of dinosaurs: the records re-examined. *Proceedings of the Geologists' Association*, 113: 185-197.

- Evans, M. (2010). The roles played by museums, collections and collectors in the early history of reptile palaeontology. Pp. 5-30 in Moody, R. T., Buffetaut, E., Naish, D., & Martill, D. M. (eds.). *Dinosaurs and Other Extinct Saurians: A Historical Perspective*. London: Geological Society.

- Fischer von Waldheim, J. G. 1813. *Zoognosia tabulis synopticus illustrata, in usum praelectionum Academiae Imperalis Medico-Chirurgicae Mosquenis edita*. 3rd edition, volume 1. 466 pages.

- Frey, E., & Tischlinger, H. (2012). The Late Jurassic pterosaur *Rhamphorhynchus*, a frequent victim of the ganoid fish *Aspidorhynchus*? *PloS One*, 7(3): e31945.

- Frey, E., Tischlinger, H., Buchy, M. C., & Martill, D. M. (2003). New specimens of *Pterosauria* (*Reptilia*) with soft parts with implications for pterosaurian anatomy and locomotion. *Geological Society, London, Special Publications*, 217(1): 233-266.

- Halstead L. B. (1975). *The Evolution and Ecology of the Dinosaurs*. Eurobook.

- Harris, J., & Carpenter, K. (1996). A large pterodactyloid from the Morrison Formation (Late Jurassic) of Garden Park, Colorado. *Neues Jahrbuch für Geologie und Paläontologie, Monatsheft*, 8: 473-484.

- Hing, Richard (2011). A re-examination of a specimen of pterosaur soft tissue from the Cretaceous Santana formation of Brazil. PhD thesis, University of Portsmouth.

- Hitchcock, E. (1858). *Ichnology of New England: a Report on the Sandstone of the Connecticut Valley Especially Its Fossil Footmarks Made to the Government of the Commonwealth of Massachusetts (No. 6)*. Boston: William White, Printer to the State.

- Hone, D. W., & Benton, M. J. (2007). An evaluation of the phylogenetic relationships of the pterosaurs among archosauromorph reptiles. *Journal of Systematic Palaeontology*, 5(4): 465-469.

- Hone, D. W. E., Tischlinger, H., Frey, E., & Röper, M. (2012). A new non-pterodactyloid pterosaur from the Late Jurassic of southern Germany. *PLoS One*, 7(7): e39312.

- Kellner, A. W., Wang, X., Tischlinger, H., Campos, D., Hone, D. W., & Meng, X. (2010). The soft tissue of *Jeholopterus* (Pterosauria, Anurognathidae, Batrachognathinae) and the structure of the pterosaur wing membrane. *Proceedings of the Royal Society B*, 277: 321-329.

- Maisch, M. W., Matzke, A. T., & Sun, G. (2004). A new dsungaripteroid pterosaur from the Lower Cretaceous of the southern Junggar Basin, northwest China. *Cretaceous Research*, 25(5): 625-634.

- Martill, D. M. (2014). *Dimorphodon* and the Reverend George Howman's noctivagous flying dragon: the earliest restoration of a pterosaur in its natural habitat. *Proceedings of the Geologists' Association*, 125(1): 120-130.

- Mayr, G., Pohl, B., Hartman, S., & Peters, D. S. (2007). The tenth skeletal specimen of *Archaeopteryx*. *Zoological Journal of the Linnean Society*, 149(1): 97-116.

- McCarthey, S. (2010). Noah's raven: who's flight of fancy? *The Guardian*, 27 September 2010. Accessed online 29 June 2014: http://www.theguardian.com/commentisfree/belief/2010/sep/27/noahs-raven-flight-fancy

- Meseguer, J., Chiappe, L. M., Sanz, J. L., Ortega, F., Andrés, A. S., Pérez-Grande, I., & Franchini, S. (2012). Lift devices in the flight of *Archaeopteryx*. *Revista española de paleontología*, 27(2): 125-130.

- Meyer, C. A., & Hunt, A. P. (1999). The first pterosaur from the Late Jurassic of Switzerland: evidence for the largest Jurassic flying animal. *Oryctos*, 2, 111-116.

- Nesbitt, S. J. (2011). The early evolution of archosaurs: relationships and the origin of major clades. *Bulletin of the American Museum of Natural History*, 352: 1-292.

- Newman, E. (1843). *The System of Nature* (2nd edition). London: Van Voorst.

- Oppel, A. (1862): Über Fährten im lithographischen Schiefer. Palaeont. *Mitt. Mus. königl. Bayer. Staates*, 1: 121-125.

- Ősi, A., Prondvai, E., & Géczy, B. (2010). The history of Late Jurassic pterosaurs housed in Hungarian collections and the revision of the holotype of *Pterodactylus micronyx* Meyer 1856 (a 'Pester Exemplar'). Pp. 277-286 in Moody, R. T., Buffetaut, E., Naish, D., & Martill, D. M. (eds.). *Dinosaurs and Other Extinct Saurians: A Historical Perspective*. Geological Society, London.

- Ostrom, J.H. (1978). The osteology of *Compsognathus longipes*. *Zitteliana*, 4: 73–118.

- Padian, K. (1983). A functional analysis of flying and walking in pterosaurs. *Paleobiology*, 9: 218-239.

- Parsons, W. L., & Parsons, K. M. (2009). Further descriptions of the osteology of *Deinonychus antirrhopus* (Saurischia, Theropoda). *Bulletin of the Buffalo Society of Natural Sciences*, 38: 43-54.

- Paul, G. S. (2002). *Dinosaurs of the Air: The Evolution and Loss of Flight in Dinosaurs and Birds*. Johns Hopkins University Press.

- Rafinesque, C. S. (1815). *Analyse de la nature, ou tableau de l'universe et des corps organises*. L'Imprimerie de Jean Barravecchia, Palermo, Italy, 224 pp.

- Rauhut, O. W. (2012). Ein "Rhamphodactylus" aus der Mörnsheim-Formation von Mühlheim. *Freunde der Bayerischen Staatssammlung für Paläontologie und Historische Geologie e.V., Jahresbericht und Mitteilungen* 01/2012; 40: 69-74.

- Reisdorf, A. G., and Wuttke, M. (2012). Re-evaluating Moodie's opisthotonic-posture hypothesis in fossil vertebrates. Part I: Reptiles - The taphonomy of the bipedal dinosaurs *Compsognathus longipes* and *Juravenator starki* from the Solnhofen Archipelago (Jurassic, Germany). *Palaeobiodiversity and Palaeoenvironments*, 92(1): 119-168.

- Rauhut, O. W., Heyng, A. M., López-Arbarello, A., & Hecker, A. (2012). A new rhynchocephalian from the Late Jurassic of Germany with a dentition that is unique amongst tetrapods. *PloS One*, 7(10): e46839.

- Schweigert, G. (2007). Ammonite biostratigraphy as a tool for dating Upper Jurassic lithographic limestones from South Germany - first results and open questions. *Neues Jahrbuch für Geologie und Paläontologie-Abhandlungen*, 245(1): 117-125.

- Seeley, H. G. (1870). *The Ornithosauria: An elementary study of the bones of pterodactyles, made from fossil remains found in the Cambridge Upper Greensand, and arranged in the Woodwardian Museum of the University of Cambridge*. Deighton, Bell.

- Sömmerring, S. T. (1812). Über einen *Ornithocephalus* oder über das unbekannten Thier der Vorwelt, dessen Fossiles Gerippe Collini im 5. Bande der Actorum Academiae Theodoro-Palatinae nebst einer Abbildung in natürlicher Grösse im Jahre 1784 beschrieb, und welches Gerippe sich gegenwärtig in der Naturalien-Sammlung der königlichen Akademie der Wissenschaften zu München befindet, Denkschriften der königlichen bayerischen Akademie der Wissenschaften, München: mathematisch-physikalische Classe 3: 89–158.

- Standish, F. H. (1821). *The life of Voltaire: with interesting particulars respecting his death, and anecdotes and characters of his contemporaries*. Andrews.

- Switek, B. (2010). Huxley and the reptile to bird transition. Pp. 249-263 in Moody, R. T., Buffetaut, E., Naish, D., & Martill, D. M. (eds.). *Dinosaurs and Other Extinct Saurians: A Historical Perspective*. London: Geological Society.

- Switek, B. (2014). The Urvogel's Old, New Clothes. *Laelaps*, 2 July 2014. Accessed online 4 July 2014: http://phenomena.nationalgeographic.com/2014/07/02/the-urvogels-old-new-clothes/

- Taquet, P., & Padian, K. (2004). The earliest known restoration of a pterosaur and the philosophical origins of Cuvier's *Ossemens Fossiles*. *Comptes Rendus Palevol*, 3(2): 157-175.

- Tischlinger, H. (2002). Der Eichstätter *Archaeopteryx* im langwelligen UV-Licht. [The Eichstätt specimen of *Archaeopteryx* under longwave ultraviolet light]. *Archaeopteryx*, 20: 21–38.

- Unwin, D. M. (2003). Smart-winged pterosaurs. *Nature*, 425: 910-911.

- Unwin, D. M. (2006). *The Pterosaurs: From Deep Time*. Pi Press.

- Wagner, J. A. (1851). *Beschreibung einer neuen Art von Ornithocephalus: Nebst krit. Vergl. d. in dk palaeontologischen Sammlung zu München aufgestellten Arten aus dieser Gattung (Vol. 3)*. Verlag d. Akad..

- Wanderer, K. (1908). *Rhamphorhynchus gemmingi* H. V. Meyer, ein Exemplar mit teilweise erhaltener Flughaut aus dem kgl. *Mineralog.-Geol. Museum zu Dresden. Palaeontographica*, 55: 195-216.

- Wellnhofer, P. (2009). Archaeopteryx: *The Icon of Evolution*. Munich: Pfeil. In English, Translated by Frank Haase.

- Witton, M. P. (2013). *Pterosaurs*. Princeton University Press.

Index

Aegisaurus leptospondylus, 15
Agassiz, Louis, 40
Anchiornis huxleyi, 97
Anurognathus, 71, *80*, 82;
    A. ammoni, 65, **80**, *81*, 82
Archaeopteryx, 2, 6, *15*, 96, 97, 98, 100;
    A. lithographica, 22, **90**, *91*, 92, 93;
    A. siemensii, 20, *91*, **94**, *95*, 96, *97*, *98*;
    classification of, 2, 22, 96
    flight, 97, 98
    soft tissue, 97
Archaeornis, 96
Ardeadactylus, 70;
    A. longicollum, 44, **70**, *71*, 83
Aspidorhynchus, 59, 100
Aurorazhdarcho,
    A. micronyx, 50;
    A. primordius, 50, 82
Batrachognathus, 82
Bavaria, 4, 8, 36, 52, 80, 86, 93;
    State Collection for Paleontology and Historical Geology, 86;
    Academy of Sciences, 94
Bavarisaurus macrodactylus, 88
Bellubrunnus rothgaengeri, 60, 83
Bennett, S. Christopher, 30, 42, 43, 50, 52, 76
Berlin, Humboldt Museum, *8*, 96
    specimen of *Archaeopteryx*, 96, 97, *99*
birds, 37, 38, 43, 47, 59, 62, 65, 84, 86, 88, 90, 92, 93, 94, 96, 98;

crown group, 2, 3, 21, 22, 26, 42, 84;
protobirds, 92, 97
Blumenbach, Johann Friedrich, 37, 52
Bonaparte, Napoleon, 35, 36
Born, Ignaz von, 20, 48, 50
Buckland, William, 40,
    Bridgewater Treatise, 40
Burian, Zdeněk, 41
cabinets of curiosity, see "nature cabinets"
Collini, Cosimo, *29*, *32*, 33, 34, 35, 36, 37
Compsognathus, 22, *86*, 88, 89, 90;
    C. coralestris, 88;
    C. longipes, 22, **86**, *87*, 88, 89;
    soft tissue, 88
Cope, Edward Drinker, 22
crocodilians, 50, 51, 54, 80;
    relationships of, 3, 27, 49
Crocodilus "multidens", 51
crown group, 3
Crystal Palace sculptures, 41
Ctenochasma, 51, *52*;
    C. brevirostris, 54;
    C. elegans, 22, 29, 38, **52**, *53*;
    soft tissue, 54
Cuvier, Georges, 21, 29, 35, 36, 37, 38, 40, 41, 47, 56
Cycnorhamphus, 70, 72, 74, 100;
    C. suevicus, **72**, *73*, 74;
    soft tissue, 74. 75
Cymatophlebia longialata, 9

Daiting, Germany, 6, 69
"Daitingopterus", 69
Dames, Wilhelm, 96
Darwin, Charles, 22, 93;
    On the Origin of Species, 93
Deinonychus, 98
Denis, Madame, 33
Diopecephalus, 44
Döderlein, Ludwig Heinrich Philipp, 80, 82
Dörr, Johann, 94
Eichstätt, town, *6*, *7*, *29*, *32*, *47*, *74*, 94;
    specimen of *Archaeopteryx*, 97
Ferdinand, Count John Freiderich, 32
Gallodactylus canjuresensis, 74
Gegenbauer, Karl, 22
Germanodactylus, 38, 66, 68, 76, 78;
    G. cristatus, 68, 69, **76**, 78;
    G. rhamphastinus, 44, **66**, *67*, 68, 69, 70, 78;
    soft tissue, 76
gingko, 17
Gnathosaurus subulatus, 20, 21, **48**, *49*, 50, 51, 54, 70,
    82, 100
Goldberg, Germany, 6
Goldfuss, Georg August, 38, 41, 58, 62, 64, 65, 66
Häberlein, Ernst Otto, 94
Häberlein, Karl, 93
Hadrosaurus foulkii, 86
Hennig, Willi, 2
Hermann, Johann, 35, 36
Hienheim, Germany, 6
Hitchcock, Edward, 90, 91
Huxley, Thomas Henry, 22, 88
ichthyosaurs, 15
International Code/Comittee of/on Zoological
    Nomenclature (ICZN), 47, 61, 68
invertebrate, fossils, 8, *9*, *10*, 15, 17;
    trackways, 8, *10-11*
Jeholopterus, 82
Jianchangnathus robustus, 65
Juravenator starki, 85
Kelheim, Germany, 6, 86, 92
lagerstatten, 4
Langenaltheim, Germany, 6, *9*, 93
Leidy, Joseph, 86
Lhwyd, Edward, 21
Linnæus, Carl, 1, *2*;
    Systema Naturae, 1
Linnaean classification, 2, 47
lithographic limestone, 4, 6, 7, 18, 32,
lithography, 4, 7
London, museum, 93, 94
    specimen of *Archaeopteryx*, *90*, 93, 94, 96
Mannheim, town, 32, *34*, 35;
    specimen of *Pterodactylus*, 21, 29, *30*, 33, 37,
    48
Maria Anna, Archduchess, 20, 48, 50
Mesolimulus walchi, *10-11*
Meyer, Christian Erich Hermann von, 50, 51, 58,
    60, 92
Moernsheim Formation, 6, *44*, *56*, 60, *66*, 69, *70*, 83
Moll, Baron Johann von, 36, 37
Mörnsheim Formation, see "Moernsheim"
Münster, Georg zu, 58
nature cabinets, 33, 52;
    at Mannheim Palace, 32, 35, 36
Niemeyer, Jakob, 94
Newman, Edward, *39*, 40
Noah's raven, 90
Nusplingen Formation, 60, 72, 74
Oberndorfer, Joseph, 86
Odontorhynchus, 56
Ornithocephalus, 37, 47, 58;

O. antiquus, 37; *O. brevirostris*, 39, 52, *53*;
    O. giganteus, 56;
    O. muensteri, 58;
    O. rhamphastinus, 68
Ornithocheirus, 41
ornithoid marsupialoids, 92
Ornithopterus, *56*, 60, 61
Ostrom, John, 93
Owen, Sir Richard, 41, 93
Padian, Kevin, 42
Painten, town, 7, 11;
    formation, 6, *72*, 74, 83, 86
Painten Pelican, *73*, 74
Painten pro-pterodactyloid, *81*, 83
Pest, Royal University of, 50;
    specimen of *Pterodactylus micronyx*, 20, 21, 48, 50
Pester Exemplar, see "Pest specimen"
phylogenetic classification, 2, 3, 69
Pfalzpaint, Germany, 6
Pleininger, Felix, 76
plesiosaurs, 15
Plectronites belemnitum, 21
Pleurosaurus goldfussi, 15
Plot, Robert, 21
Pohl, Burkhard, 97
Pterodactylus, 41, 47, 48, 58, 60, 62, 64, 69, 70, 76, 80, 93;
    classification, 38-41, 44;
    P. antiquus, *21*, 22, *30*, 32, *34*, 37, *38*, *39*, *40*, 41, 42, 43, **44**, *45*, 46, 52, 60, 68, 69, 76, 83;
    P. crassipes, 93;
    P. crassirostris, 38;
    P. dubious, 68;
    P. elegans, 54;
    P. eurychirus, 74;
    P. kochi, 76;
    P. longicaudus, 58, 61;
    P. longicollum, 70;
    P. longipes, 70;
    P. medius, 38, 66, *67*, 68, 69;
    P. micronyx, 21, 50, 76;
    P. spectabilis, *40*;
    P. suevicus, 72;
    P. wuerttembergicus, 72;
    soft tissue, 44, 46
Pteordaustro, 51
Pteromonodactylus, 56
pterosaurs, 27, 29, 43;
    anatomy, 38, 42, 76;
    classification, 29, 30, 40;
    early art, 34, 35, *40*, 41, 42;
    early fossils, *20*, 32, 41, 56;
    definition, 29;
    growth, 32;
    soft tissue, 41, 42;
    diversity, 30, 64, 80, 82, 83
Pterotherium, 47
reef, 15, *17*;
    coral, 8, 9, 11, 15, *16*;
    fish, 15;
    production, 15;
    sponge, 8, 9, 11, 15
Rhamphorhynchus, *15*, 41, 56, 58, 59, 60, 61, 65, 80, 83, 100;
    fatal collisions with fish, 59;
    R. carnegiei, 82;
    R. langicaudus, 58;
    R. muensteri, 22, *29*, *40*, **56**, *57*, 58, 59, 60, 64, 70;
"Rhamphodactylus", 83
Ried, Germany, 6

Rögling Formation, see "Painten, formation"
Rutellum impicatum, 21
Scaphognathus, 38, 41, 46, 62, 64, 65, 66, 71;
    S. crassirostris, *31, 39, 40*, **62**, *63*, 64, 65;
    soft tissue, 64
Scrotum humanum, 21
Seeley, Harry Grover, 64, 66, 72
Senefelder, Alois, 7
sharks, 15
Solnhofen, archipelago, 6, 13, 17, 71, 100;
    town of 6, *7*;
    limestone, 4, 6, *8*, 11, 30, 65, 82, 100;
    fauna, 6, 15, 30, 76, 82, 83, 86;
    flora, 17;
    formation, 6, 7, *17*, 20, 21, 22, *44, 52*, 50, 56, 58, 59, 60, *62*, 65, *66*, *70*, 72, 74, *76*, 78, 80, 83, 84, 86, *90*, 94;
    lagoon, 9, 11, 15, 17,
    specimen of *Archaeopteryx*, *92*;
    topogrphy, 15, *16*, 17
Sömmerring, Samuel Thomas von, 37, 38, 41, 52
Sordes pilosus, 41
Southern Franconian Alb, 4, 7, 11, 15, 83, 86
stem-birds, 3, 6, 17, 18, 20, 21, 22, 26, 27, 30, 42, 43, 58, 84, 88, 98, 100;
    diversity, 3, 6, 22, 26, 100;
    earliest finds, 21, 90, 92
stem-crocodilians, 15, 17, 51
stem-tuatara, 15, 17
stem group, 3, 29
teleosaurids, 51
Tethys Sea, 8, 11, 13, 15
Theodore, Elector Karl, 32, 33, 36
Thermopolis specimen of *Archaeopteryx*, *20*, 96
theropods, 84, 88
typological classification, 1, 2

Verhelst, Egid, *29*, 34
Voltaire, 33
Wagler, Johann Georg, 38
Wagner, Andreas, 68
Wellnhofer, Peter, 30, 42, 50, 51, 68, 76
Workerszell, Germany, 6
Young, C.C., 76
Zandt, Germany, 6

Image Credits

The following photographs have been released under a Creative Commons Attribution Share-Alike license (CC-BY-SA 3.0):

By Masur: p. 8
By Dalibri: p. 9a
By Dr. Alexander Meyer: p. 9b
By Ghedo: p. 11; p. 71
By Stephan Shulz: p. 19
By Hubert Berberich: p. 34-35
By Nachosan: p. 58
By Alex Giltjes: p. 74
By H. Zell: p. 78
By Ballista: p. 89
By H. Raab: p. 90; p. 97; p. 99
By Vitold Muratov: p. 92

For more information and specific terms of re-use, see creativecommons.org

Historical paintings and illustrations produced prior to the 20th century are reproduced from the Public Domain.

All other images are copyright Matthew P. Martyniuk 2014, all rights reserved.

Printed in Great Britain
by Amazon